LADY LUCK

The Theory of Probability

Warren Weaver

Dover Publications, Inc.
New York

Illustrations by Peg Hosford

Several illustrations were redrawn with the permission of the following: FIGURE 11, John Riordan, *An Introduction to Combinatorial Analysis* (New York: John Wiley & Sons, Inc., 1958). FIGURE 12, Kemeny, Snell, and Thompson, *Introduction to Finite Mathematics* (Englewood Cliffs, New Jersey: Prentice-Hall, Inc., 1957). FIGURES 31, 33, 36, 41, Mosteller, Rourke, and Thomas, *Probability and Statistics* (Reading, Massachusetts: Addison-Wesley Publishing Company, Inc., 1961).

Published in Canada by General Publishing Company, Ltd., 30 Lesmill Road, Don Mills, Toronto, Ontario.
Published in the United Kingdom by Constable and Company, Ltd.

This Dover edition, first published in 1982, is a slightly revised republication of the work originally published in 1963 by Anchor Books, Doubleday & Company, Inc., Garden City, New York, as part of the Science Study Series. Omitted from this edition are the biographical sketch of the author, the general preface to the Science Study Series and the listing of that series.

Manufactured in the United States of America
Dover Publications, Inc.
31 East 2nd Street
Mineola, N.Y. 11501

Library of Congress Cataloging in Publication Data

Weaver, Warren, 1894-1978
Lady Luck : the theory of probability.

Reprint. Originally published: Garden City, N. Y. : Anchor Books, 1963. (Science study series)
Bibliography: p.
Includes index.
1. Probabilities. I. Title. II. Series: Science study series.
QA273.W4 1982 519.2 82-7396
ISBN 0-486-24342-7 AACR2

She doesn't like mathematics, but
she is awfully good at probabilities.
So this is for Mary.

The real trouble with this world of ours is not that it is an unreasonable world, nor even that it is a reasonable one. The commonest kind of trouble is that it is nearly reasonable, but not quite.

G. K. CHESTERTON, "Orthodoxy"

FOREWORD

You, the reader, have a right to know why I wrote this little book and what I am trying to do in it. Particularly, you should be warned at the outset (I hope before you pay your money over) what I do not claim, either for the book or for myself.

Forty years ago, and for a decade thereafter, I was a professor of mathematics at the University of Wisconsin. On several occasions I taught a course in the mathematical theory of probability. By all present-day standards it was a very elementary course, which used essentially no mathematical concept or technique not taught in a good solid course in calculus.

In 1932 I stopped being a mathematician in any professional sense, and then as an officer of the Rockefeller Foundation I had a wonderful chance to view scientific activities all over the world, particularly in biology, agriculture, and medicine, these being fields in which I had not been trained. This may have enlarged my field of view, but it tended to blur my focus as regards technical details. I kept on reading a good bit of mathematical literature, simply because I love the subject and enjoyed the reading.

During World War II, Dr. Vannevar Bush asked me to take charge of an Applied Mathematics Panel of the Office of Scientific Research and Development, which he headed. This panel had the duty of bringing mathematics usefully to bear on a wide range of defense (and offense) problems. It turned out that a large proportion of these problems involved probabilistic or (which is closely related) statistical reasoning. This association put me into intimate, eighteen-hours-a-day contact with a large number of our country's mathematicians — many of them experts in mathematical statistics, some of them probabilists, many more of them mathematicians, applied or "pure," who were eager to bring their abilities to the defense of freedom.

The experience deeply reinforced a conviction I had always had — that the type of thinking about problems which one learns in probability theory is more than interesting, is more than fun (although I count that very valuable); it is of the highest importance. For no other type of thinking can deal with many of the problems of the modern world.

So I have written this modest little book. I say "modest" with no affectation whatsoever. It is not a profound book, for the author is unfortunately not a profound scholar. It is not very "modern"; for it only hints at the beautiful developments of the last quarter century.

I have written it for a double audience: first, for bright high school youngsters, in the hope of attracting them to the field and to more serious further study; and second, for *people*. By people I guess I mean any and all sorts of adults, in the hope that they will be amused, that they will be intellectually stimulated, and that they will be sufficiently interested to use their influence — with Parent-Teacher groups, school boards, etc. — to see to it that this rich and practical and lovely subject gets some reasonable amount of attention in the mathematical courses in high schools. Indeed, I am convinced that this way of thinking ought to start in elementary schools.

So don't expect the systematic development one would find in a textbook. If you are seriously enough interested in this subject to be willing to subject yourself to the discipline of a textbook, then by all means get hold of *Probability and Statistics*, by Mosteller, Rourke, and Thomas (Reading, Massachusetts: Addison-Wesley Publishing Company, 1961), this being the volume prepared for the outstandingly excellent TV presentation on Continental Classroom.

You will find that the book you now have in your hand does not settle down to its direct subject matter until Chapter II. If you are impatient to start reading about probability theory, you had better skip the first chapter for the time being. If you are interested in having some background as to how probability reasoning fits into the general problem of thinking, then you will want to read this chapter. But if you do read it, don't be discouraged. The rest of the book is less abstract and less general in tone.

As you read along, don't expect to find a philosophical treatise. Don't expect meticulous accuracy about the very latest development in pure mathematics. Just bump along with me, admittedly a little unevenly, admittedly on side roads now and then, and see if it isn't really fun.

Warren Weaver

CONTENTS

Because of the omission of material that was no longer relevant from the original edition (see copyright statement for details), the present volume skips the page numbers 11 through 20. No part of the basic text and its supplements has been omitted.

Chapter I

THOUGHTS ABOUT THINKING

Some billion years ago, an anonymous speck of protoplasm protruded the first primitive pseudopodium into the primeval slime, and perhaps the first state of uncertainty occurred.

I. J. GOOD, *Science*, Feb. 20, 1959

The Reasoning Animal

This book is, in one sense, about thinking. About a certain way of thinking, that is.

You may have the idea that it is dull to think about thinking, and perhaps you even have the idea

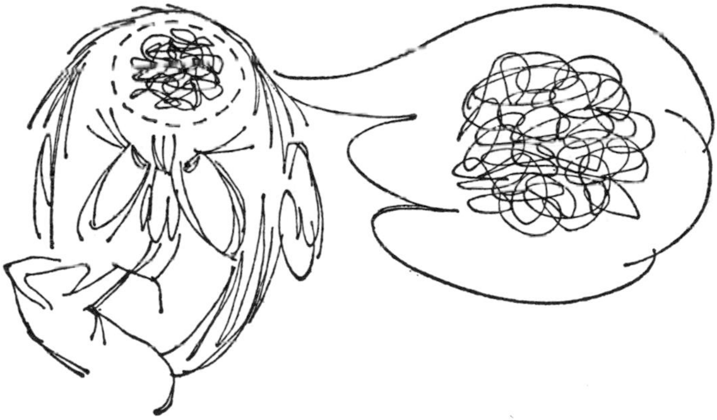

FIG. 1

that it is unprofitable or too difficult to improve the way you think. But, you see, you had to think to come to those conclusions. And how certain are you that your conclusions are sound?

You may, for the moment at least, be rather bored at the idea of thinking about thinking; but the completely obvious fact is that every person, young or old, is continuously having to face questions, is continuously having to "figure things out," is continuously, in short, *having to think.*

I say that you may at the moment be almost bored at the prospect of thinking about thinking. But this book is going to introduce you to a special way of thinking, a special brand of reasoning which, I am confident, you will find not only useful, but fun as well. It will be about a type of thinking that, when stated boldly, seems a little strange. For we often suppose that we *think* with the purpose of coming to definite and sure conclusions. This book, on the contrary, deals with thinking about uncertainty.

The ability to think out answers to new problems is a proud characteristic, almost a defining characteristic, of human beings. People are not always reasonable — that's for sure. But people *do* reason. They do try to figure things out.

Animals appear to figure out the answers to certain rather simple questions. To be really fair to them, animals do succeed with some pretty complicated and queer problems. Honey bees can keep track of directions by using the sun, and they use the state of polarization of the light to tell them where the sun is when it is cloudy. Salmon find their way back to the streams where they were born, very possibly identifying their native location because it *smells* to them like home. (Perhaps you didn't realize that fish have any sense of smell!) Birds navigate accurately over hundreds of miles of open sea, and, according to one theory, can use the stars for guiding them at night. It was recently reported that five penguins were taken by airplane from their home at Wilkes Station to McMurdo Sound, in Antarctica. When they were released, three of them waddled more than 2000 miles across the bleak and monotonous ice to return to their native rookery. Figure that one out — and remember that the penguins solved the problem! And we have all been reading, recently, about the remarkable intelligence of porpoises.

Mice learn from experience what turns to make in order to "run" a maze and to end up at the place where food will reward them. There is considerable evidence that learning, or at least something closely related to learning, has been demonstrated for certain small, chiefly aquatic, worms called *Planaria* or flatworms. Professor James V. McConnell, of the Psychology Department of the University of Michigan, has become a sort of impresario for these lowly and timid performers, and has stimulated dozens of other investigators, and many high school students, to perform experiments to see to what extent these worms can "learn" not to do something which leads to punishment, or can "learn" to crawl simple mazes in order to obtain rewards. The "Planaria Research Group" at Michigan issue, quarterly, a lively mimeographed collection of reports and letters under the title *The Worm Runner's Digest.*

In fact, something at least very much like learning

was demonstrated nearly a half-century ago for still more primitive organisms. The great biologist H. S. Jennings carried out a series of experiments[1] on a one-celled, nearly transparent protozoan called *Stentor roeselii.* This little creature, one-hundredth to one-fiftieth of an inch in length, is shaped a little like the flower of an Easter lily, turned upside down, and is found attached to water plants in marshy pools. Summarizing his observations, Dr. Jennings said, "The organism 'tries' one method of action; if this fails, it tries another, till one succeeds"; and he concluded, "The phenomena are thus similar to those shown in the 'learning' of higher organisms, save that the modifications depend upon less complex reactions and last a shorter time."

Just *how* animals figure such things out is very hard to say. It is sometimes difficult to distinguish between *instinctive* behavior, which is a pattern of action creatures are born with, and *learned* behavior, which involves some type of thinking. One obvious difficulty in deciding how much animals can think is that they cannot directly tell us, at least in the unambiguous terms of our language, how they do things.

We are just beginning to discover something about the communication systems which certain animals use. You ought to read what Professor Karl von Frisch has learned about the way honey bees convey messages to each other by the patterns in which they dance inside the hive. I suppose rock-and-rollers (and ballet dancers) will reply that human beings communicate through dance also!

[1]H. S. Jennings, *Behavior of the Lower Organisms* (New York: Columbia University Press, 1906).

Communications from man to some animals can be rather highly developed, as anyone with a well-trained dog knows. But communication from animals to man is more limited. Your dog can let

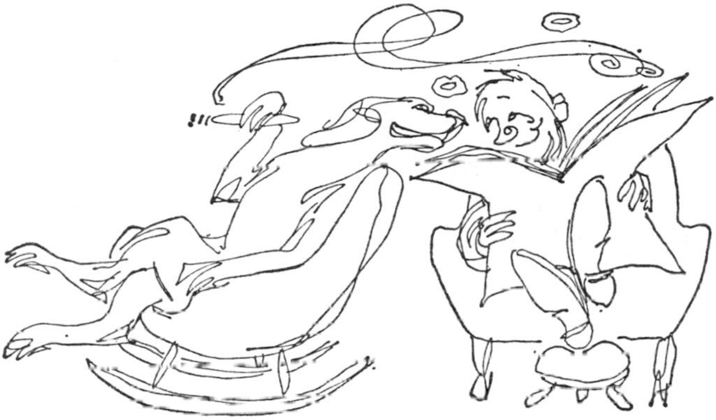

FIG. 2

you know when he is hungry, or when he wants to go outdoors. He can warn you when a stranger approaches. He can let you know, without any doubt at all, that he trusts you and likes you. But he can't carry on any general conversation with you.

Yes, the human species, at least at our present state of knowledge of the rest of the animal world, can be defined as the special animals who reason, who can freely communicate their thoughts by speech, and who can record their thoughts in the symbolic form we call writing.

Reasoning and Fun

Reasoning gives a person a satisfying sense of accomplishment, especially when it leads to the clearing up of a puzzling curiosity. And it can be great fun. It is often in games and puzzles that we first experience the fact that reasoning *is* fun, particularly when it leads to a conclusion that is interesting and one that was unsuspected until the reasoning was carried through to a conclusion.

It is a great pity that many people develop the idea that thinking is painful or dull. This opinion is even, irrationally, held by persons who enjoy working crossword puzzles and Double-Crostics. Some go so far as to consider thinking an impractical activity, indulged in only by "highbrows."

But no person can afford to be disinterested in thinking and in the orderly kind of thinking which we usually call reasoning. Life is a pretty confusing business for us all — for young people, for the businessman, for the housewife, and for the older persons who, although they may not have to make active decisions every hour of every day, now have the time to wonder "what it is all about."

Yes, young or old, clever or average, we all face questions that require answering, problems that must be solved, decisions that must be made; and it is important to realize that the processes of analysis can be pleasant, rather than painful.

The Kind of Questions We Have to Answer

In only a very, very few instances do we ever, in real life, meet problems that have just one single certain answer. In arithmetic, to be sure, two times one is exactly 2.0 — not 2.1, or 1.9, or something else. But that result, we must remember, holds good only when we are dealing with objects which obey the assumptions on which arithmetic is based. Two simultaneous death penalties are not twice as severe as one, and the reader can easily amuse himself or herself by thinking of other instances in which the ordinary rules of arithmetic do not apply. For example, is A + B always exactly equivalent to B + A? In ordinary arithmetic the two are identical; but if you are making a cake, will mixing plus baking produce the same result as baking plus mixing? And can't you think of instances in which one plus one is not equal to two?

Usually — in games, in business, in politics, in love affairs, in forecasting the weather or the stock market, and (as we will see in this book) in all the basic phenomena of biology and chemistry and physics — the questions which really bother us are both more vague and more complicated than they are in arithmetic or geometry. And very seldom do they have just one clear-cut and certain answer.

In the vast majority of instances, we never have *enough* knowledge or *certain enough* knowledge to permit anyone, even the professional expert on reasoning, to come to one single definite and certain conclusion.

Had I better carry an umbrella today? What chance has Notre Dame of winning the big game next Saturday? If we raise the college tuition by $300, will it discourage so many applicants that we will end up with *less* income from fees? How seriously should our government take Khrushchev's threat that he will "protect" Cuba with rockets? Is hospital insurance a good bet for me? Isn't it a wonderful idea to buy a two-dollar ticket at the race track every day on that particular daily double which would pay the greatest amount? Doctor, if I have the left lobe of my lung removed, what is the chance that the cancer will really be cured? I heard somewhere that red once came up at Monte Carlo twenty-nine times in a row: Is that possible? What about those two bright young men who broke the bank at Monte Carlo? How much is it sensible for me to spend on a lightning protection system for my new house? I know that the polls give the Democrats a small edge, but what does that mean? The only evidence linking lung cancer and cigarette smoking is *statistical,* they tell me — and isn't all this statistical evidence just plain hooey? If I run a corner grocery and sell an average of two boxes of Superduper Soap Flakes a day, how large a stock should I maintain in order to run a small risk — say one in ten — of disappointing a customer? Does it make sense for our company to spend a half-million dollars drilling for oil in this particular location? Pete Kelly has a batting average of .286 and hasn't had a hit the last five times at bat: According to the law of averages, isn't he due for a hit?

What Kind of Reasoning Is Able to Furnish Useful Replies to Questions of This Sort?

This book will introduce you to that kind of reasoning. And in the closing chapters there will be an indication (necessarily rather brief and without technical details) of the fact that modern science now recognizes that it is this kind of reasoning — *probabilistic reasoning* — which explains the truly basic happenings in the physical universe in which we exist, and a large part of the happenings in the world of living things.

The almost unlimited usefulness of probabilistic reasoning is one of the most important and striking aspects of modern science. This viewpoint has been developed and accepted chiefly over the last half century, although we will learn that probability theory originated more than three hundred years ago.

After the present introductory chapter, you are going to read the story of the origin of probability theory. It is a fascinating and very human tale. But even before we start with the original history, we ought to back off and get a long view, to give perspective to the later details.

Thinking and Reasoning

Why did we refer, a few paragraphs back, to *probabilistic* reasoning? What sort of reasoning is that? How does probabilistic reasoning differ from the other older and classical forms of reasoning? And why have we part of the time been speaking of *thinking*, and part of the time of *reasoning?*

If you go to a large library and look up the subject of thinking, you will find some pretty difficult material. "When older philosophers began to think about thinking, and how by thinking we reached truth, they commonly found themselves writing very long books, very hard to read." The author of this quotation, John Dewey, himself wrote a small and attractive book, *How We Think.*

There are two somewhat strange uses of the word "think" which we will mention just to make it clear that we are not at all concerned here with those two special varieties of thinking. In one of these uses "I think" and "I imagine" are about synonymous. For example, when a child says, "I like to think about fairies," he is using the word think in this particular sense.

In a second special meaning "I think" is very closely equivalent to "I believe," especially if the belief is an unfounded one, a prejudgment, or the result of prejudice. Thus the closed-minded superpatriot declares, "I think there is absolutely nothing good about the Russian social and political system."

There is still another kind of thinking with which this book will have no concern. That is the type often called "reverie." This is the largely automatic, unregulated, more or less uncontrolled sequence of recollections, vague notions, apprehensions, hopes, etc., which form an undercurrent to more systematic thinking. In a specially unorganized and uncontrolled fashion, this reverie process goes on in dreams, when we are asleep.

Sometimes when, for a time, we think hard about some subject and then "dismiss it from our minds," this subject seems to reappear later, in an improved and expanded form, just as though it had descended to a subconscious level of the mind, and had there profited by association with an unrecognized mixture of other ideas in a sort of unconscious reverie. I think writers often experience this. If they have to write about some topic, they may have to wait a few days, or weeks, or even months, while the topic develops itself at an almost completely subconscious level. Then suddenly, the author finds himself ready to write page after page, almost without effort.

I have said we are not going to be concerned here with purely imaginative thinking, nor with believing-thinking, nor with reverie. We are concerned in this book with the reflective type of thinking which is ordinarily called *reasoning*, in which one step leads in an orderly way to the next step, the whole consecutive process drawing finally to a conclusion.

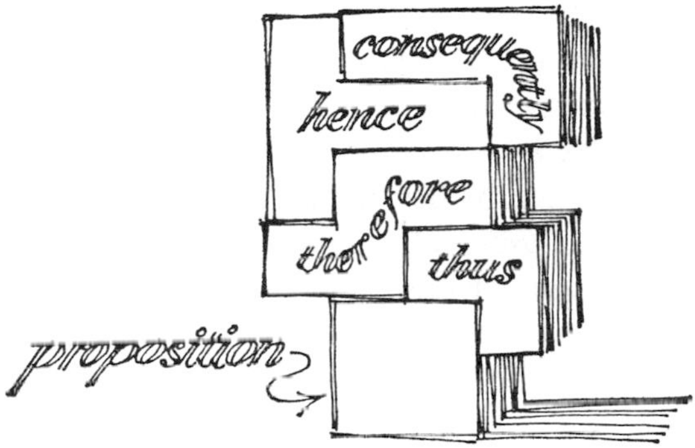

FIG. 3

This kind of thinking reaches a particularly refined level in mathematics. Its precision in mathematics depends upon several things — the unambiguous way in which terms are defined, the restraint with which the definitions are obeyed, and the care with which all the rules of procedure are set out and made clear.

This kind of thinking is *logical* thinking; and logic itself can be defined as the systematic study of the conditions and procedures which permit *valid inference* — that is to say, which permit one to start with one or more statements (or "propositions" as the experts like to say) and derive from these, or *infer* from these, one or more *new* statements or propositions which are *valid* in the sense that they are justified by, and are in strict fact consequences of, the starting statements.

It is of the highest importance to recognize that logic does not make something from nothing, that it merely unrolls out to view, so to speak, statements and propositions and relationships which were present in, although hidden within, the statements used as the starting point.

Thus what is important about a logical inference is not its *truth*, but rather its *validity*. A logical conclusion can properly merit the adjectives *correct*, *sound*, or *accurate*, all meaning that it has been dependably and properly derived from the starting material. But the fact that it has been produced by correct logical methods is of itself not the slightest evidence that it is *true*. If the starting statements were, in some way, known to be true, then the logically derived consequence would also be true. But if you are interested in truth, then either you have to go back and establish the truth of the starting statements, or, neglecting the logical process which has produced them, you have to establish the truth of the inferred statements by some method directly supplied to them.

Bertrand Russell once declared that mathematics is the subject in which you never know what you are talking about, and never know whether or not what you say is true. What we have just said should explain the final part of his description; for we have seen that validity, not truth, is the hallmark of a logical conclusion. The earlier part of Russell's somewhat mischievous but accurate definition is, incidentally, equally sound and equally important. Mathematics is so magnificently *general* a discipline, that it is literally true that one does not know, in trivial particular, what one is talking about. Thus the mathematical identity $2x + 2x = 4x$ is valid, even though you have no idea what x stands for.

Classical Logic

The type of reasoning used in geometry and algebra is that which has been developed in classical logic. It involves three steps. First, you start out with a set of *premises* (statements, or postulates) which are, at least in that particular discussion, to be accepted without question. In modern mathematics these initial premises are recognized to be pure assumptions. Long ago mathematicians — Euclid, for example — started out with what were called "axioms," these being statements which were then presumed to be self-evidently true. But this alleged truth turned out to be a snare; for it was later found out that several of these "self-evidently true" statements could be replaced with other statements (not necessarily at all the same) and that self-consistent, but different, theories could be developed from them.[2] So nowadays the starting step consists of premises simply assumed without further discussion.

The second step involved the application, to the starting premises, of the rules of logic. The core of deductive logic is the so-called syllogism, which

[2]Thus Euclid considered it obvious that one and only one line can be drawn parallel to a given line through a point not on the given line. But it turned out that there is an important system of geometry in which *no* line can be drawn "parallel" to a given line, and another in which *more than* one "parallel" can be so drawn.

consists of a major premise, a minor premise, and a conclusion. In general terms:

Major Premise:	Every M is P,
Minor Premise:	S is M,
Conclusion:	Therefore S is P.

Or, to cite a particular example:

Every virtue is laudable,
Kindness is a virtue,
Therefore kindness is laudable.

It is really remarkable that from this kind of start, which seems so innocently simple, there has been developed a grand array of powerful procedures by means of which reasoning can be kept tidy and dependable.

Lewis Carroll, the author of *Alice in Wonderland* and a mathematician at Christ Church, Oxford, wrote several books and pamphlets explaining how to apply the rules of logic, and — not wholly leaving wonderland — he showed how one could, from the three premises

1. Babies are illogical,
2. Nobody is despised who can manage a crocodile,
3. Illogical persons are despised,

derive the conclusion:

Babies cannot manage crocodiles.

In fact, since *babies are illogical* (Premise 1) and since *illogical persons are despised* (Premise 3) it follows that *babies are despised.* And since *despised persons cannot manage crocodiles* (Premise 2), it follows that *babies cannot manage crocodiles.*

This example of logical reasoning indicates, incidentally, that a logical conclusion is worth just exactly what the premises are worth. Also you will say, and properly, that you could solve this little puzzle without studying logic. But take a look at another of Lewis Carroll's examples. What is the conclusion from the following nine interlocked premises?

1. All who neither dance on tightropes nor eat penny buns are old.
2. Pigs that are liable to giddiness are treated with respect.
3. A wise balloonist takes an umbrella with him.
4. No one ought to lunch in public who looks ridiculous and eats penny buns.
5. Young creatures who go up in balloons are liable to dizziness.
6. Fat creatures who look ridiculous may lunch in public provided they do not dance on tightropes.
7. No wise creatures dance on tightropes if liable to giddiness.
8. A pig looks ridiculous, carrying an umbrella.
9. All who do not dance on tightropes and who are treated with respect are fat.

This one takes a bit of thinking; but it is straightforward, using the procedures of logic, to discover that these nine premises imply the conclusion: *No wise young pigs go up in balloons.*

Now you may not be overwhelmed by the significance of these examples, but, on the other hand, you must admit that it is rather impressive to realize that involved and complicated premises do in fact lead unambiguously to certain definite conclusions,

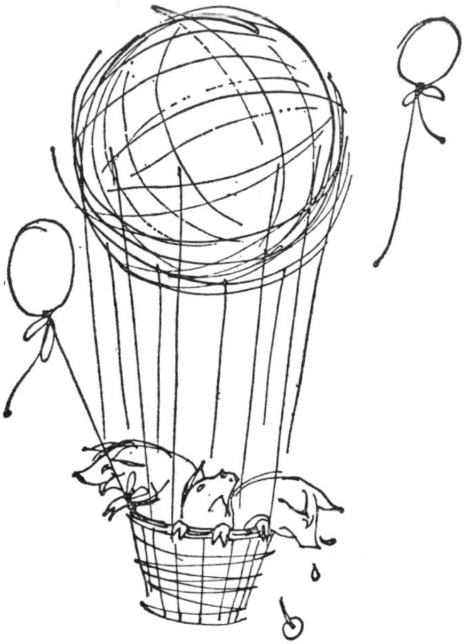

FIG. 4

and that clearly formulated procedures of logical thinking can unravel a curious mess of interlocked statements of this kind.

But the most important aspect — the third and culminating stèp of the classical procedure that one often first learns in geometry — is precisely that there *is* a definite conclusion, that with open-and-shut finality one can be sure that the conclusion is, under the given premises, completely valid, and that its opposite is completely false.

Be sure you did not miss, in the preceding sentence, the critically important clause "under the given premises." For this type of reasoning can, of course, produce nothing more than an "if-then" kind of result. That is, *if* the assumptions hold, *then* the result is valid. Indeed it is then unquestionably valid, and its opposite is unquestionably false.

One can have nothing but enthusiasm and respect for the development of the rules and procedures of logic, and a university course in logic shows how complex and fascinating and dependable they can be.

With a given start, this process enables one confidently to decide that certain statements, on the basis of the given start, are *true* or *false*. To certain questions one can confidently respond "yes" or "no."

But this process in its classical form (which in deference to its major originator is sometimes called Aristotelian logic) has an absolutely devastating limitation.

The trouble is that there are so very many interesting and important questions to which this logical method does not at all apply. There are so many questions which cannot possibly be answered with either yes or no. There are so very many cases in which our starting knowledge is simply not sufficiently extensive, or sufficiently reliable, to lead to a completely positive yes or a completely negative no. Much more often the answer should be, to use the phrase made famous by Sam Goldwyn, "a definite maybe."

In logic it is customary to speak of "truth values"; and classical Aristotelian logic is concerned with only two truth values — absolute and perfect truth, which can be symbolized by the number 1, and absolute and total falsehood, which can be symbolized by the number 0. On that basis, classical logic puts all statements into one or another of these categories. The first category contains statements which receive the score of 1, or *true*. The second contains statements which receive the score of 0, or *false*. A third, and embarrassingly large, category contains all the statements which classical logic cannot handle at all, many of these statements presumably being neither completely true nor completely false.

A great limitation of classical logic is contained in the fact that there are so very many interesting and important statements which fall in the third category with which classical logic cannot deal at all.

There are other limitations of classical logic. A well-known logician and philosopher, Susanne K. Langer, contends that the expression of ideas in words — which condemns one to a *serial* or step-by-step procedure since one necessarily writes or speaks one word after another — is too limited; and that the mind, not using words at all, is able to deal simultaneously with a whole interrelated set of ideas. Perhaps the total and unvocalized response which one has when he looks at a beautiful painting is an example of what Dr. Langer means.

One of the most fundamental and exciting intellectual feats of this century, moreover, has been the discovery by a mathematician and logician named Kurt Gödel that logic has some built-in limitations which had previously been unsuspected. When one wishes to develop any logical system (say the logical basis for arithmetic, or the logical basis for some field of physics) he has to start, as we have already seen, with a set of postulates — some assumptions. It is naturally a matter of the greatest concern to be sure that the set of postulates chosen is internally consistent so that they do not in some complicated and concealed way contradict each other. Gödel proved the absolutely stunning result (stunning in all senses) that it is impossible — actually *impossible*, not just unreasonably difficult — to prove the consistency of any set of postulates which is rich enough in content to be interesting — rich enough, that is, in the sense of leading to a useful body of results. The question, "Is there an inner flaw in this logical system?" is a question which is unanswerable!

One often hears the complimentary phrase "flawless logic"; and it is hard to realize that this discipline, so long considered the one kind of thinking or reasoning beyond criticism, should have this mysterious inner flaw, which consists of the fact that you can never discover whether or not it really has a flaw.

Gödel also proved that any deductive logical system has a further great limitation. If you could eliminate completely the third category I have described (the category of statements with which classical logic makes no pretense of dealing), then you certainly should have the right to hope and expect that logic would be able to take all the other statements — all being of the kind which is admittedly the business of classical logic — and neatly put them either into category 1 (true) or category 2 (false). But Gödel proved that this is not so! He demonstrated that it is *always* possible, within a logical system, to ask questions which are undecidable!

The last-described limitations of classical logic are of profound interest to the experts, but they need not worry the rest of us too much. Least of all do they constitute any justification for thinking illogically. But the limitation to only two truth values does in fact keep most of us, and most of the time, from getting a great deal of practical help from logic.

It is at just this point that the *logic of probability* comes to our rescue. It is capable of taking a whole series of statements, no one of which is either totally false or totally true, and of ordering these statements relative to their *degree* of truth, telling how much more or less plausible one is than another. Probability logic is not restricted to *two* truth values, 0 and 1, but makes use, as we will see, of an infinite set of truth values, expressed as numbers lying between 0 and 1.

Probability theory is capable of considering situations in which we just haven't enough information to permit the application of classical logical thinking, and it is capable of giving the best kind of answer which the incomplete evidence justifies. In a great many cases probability theory not only says, "My advice is thus and so," but also can tell what degree of confidence you are justified in placing in the advice given. The rest of this book will, I hope you agree at the end, justify this paragraph.

Probability reasoning supplements classical logical thinking, rather than displaces it. Indeed, classical logical thinking is used at every step to produce probability theory. This is a good example of the fact that science is, in the fine phrase of Dr. James B. Conant, an "accumulative" endeavor, in which each new development may refine and extend past knowledge, but is always based upon past knowledge. Science builds up like a strong wall, each stone resting solidly on the lower ones.

Although we will all have to wait until late in the book before we can appreciate the full significance of probability theory, I want, here at the outset, to give at least a hint as to why, in my judgment at least, this type of thinking is a plain necessity in the modern world.

I will, for the present, base this strong claim on a single illustration.

The problem of peace and war[3] is, all will agree, of absolutely overriding importance and urgency. One aspect of this problem, and one which profoundly influences the relations between nations, to say nothing of the internal economy of each nation, is the reduction of armaments.

As one reads the publicly available documents and statements, it is hard to avoid the impression that our government has[4] proceeded on the Aristotelian view that there are only two solutions — an absolutely right one, which is ours, and an absolutely wrong one, which is Russia's. Of even greater importance, our government has seemed to believe that we must insist on a formula of procedure (step-by-step controls, inspection, etc.) which will *guarantee* success — will guarantee success *100 per cent.*

That, I would most solemnly urge, is an outmoded Aristotelian view! Security does not have two truth values, zero and one. There is a whole array of intermediate possibilities. As long as we insist on procedures which we think will assure the value "one," we will, first of all, make little or no progress. And in the meantime, we will be running a serious risk of realizing zero security.

[3]Problems of strategy also require the application of *game theory*, which is not concerned primarily with probabilistic reasoning, but rather with the mathematical techniques which have been developed for maximizing gains and minimizing losses.

[4]At least up to the summer of 1962.

One often hears that any system of inspection of armaments must be "foolproof," and that if we cannot get a foolproof system we will accept no system at all. But there simply is no such thing as a foolproof system.

What we need is probabilistic thinking, which is for so many practical and serious problems the *only* realistic kind of thinking. We must be trained in the technique of weighing alternative risks. We must face the grim fact that we cannot eliminate risk. The problem, therefore, is to be able to evaluate risks, to identify a reasonable risk, and to adopt it with courage, and care, and hope.

Presidents and generals and politicians need probability theory. Businessmen and doctors and lawyers and engineers and housewives need probability theory. High school boys and girls perhaps need probability theory most of all for it is they who in a few years will become all those other people.

The important problems in life usually require a comparison of the odds. So let us turn back three centuries, and learn about a problem in odds.

Chapter II

THE BIRTH OF LADY LUCK

It is remarkable that a science which began with the consideration of games of chance should have become the most important object of human knowledge.

LAPLACE[1], *Théorie Analytique des Probabilités*

Most scientific subjects have grown step by step, starting from very simple notions that, as one tries to trace them further and further back in time, finally disappear in the mists of antiquity. For example, arithmetic, which in its modern refinements forms a very sophisticated body of ideas about numbers, presumably started out with the need of primitive man to count his few possessions — his cattle or perhaps his wives. He may have used his fingers; and that may have led to our modern use of ten as the base of our number system.

Astronomy, which has today become so detailed and vast a scientific field with the most elaborate instrumentation, undoubtedly originated in the earliest curiosity of men about the beauty and

[1]Pierre Simon, Marquis de Laplace (1749–1827) was a French mathematician and astronomer. His *Mécanique Céleste* is the classic treatise on the movements of celestial bodies in a Newtonian universe.

wonder of the night sky, the majesty of the daily visit of the sun, and the strange waxing and waning of the moon with its supposed influence on various human activities.

We think of molecules and atoms as being peculiarly characteristic of the most modern science. But the Roman poet Lucretius, in his famous poetic essay *On the Nature of Things*, written in the first century before Christ, said:

> Because that *Elements* can not be spi'd
> By human eyes; behold what bodies now
> In things thou canst not see, yet must allow.

And some four hundred years earlier than that, Democritus of Abdera taught that the world consisted of empty space plus an infinite number of indivisible, invisibly small "atoms."

For a lot of reasons it is clear that from very ancient times men must also have been concerned

FIG. 5

with, and curious about, probability and chance. The "casting of lots," the playing of simple games of chance, the worry and indeed the fear that were inevitable in a world of events determined by the caprices of the gods — all this assures that men were bound to think about chance and probability, and "fate," even though their reasoning necessarily would be very simple and faulty as judged by our present-day standards.

The rather strange fact is, however, that the theory of probability — the *mathematical* theory of probability — had a clearly recognizable and rather definite start. It is known just when Lady Luck was born. In fact, this event occurred in France slightly over three hundred years ago,[2] in the year 1654.

A distinguished Frenchman, Antoine Gombauld, Chevalier de Méré, Sieur de Baussay, lived the life of a gentleman of the period, dividing his time between his estate in Poitou and the court at Paris. He was admired for his philosophical skill, for the wisdom of his advice on subtle matters, and for his charming personality. He undoubtedly dropped in at the gaming rooms from time to time, and he clearly had an inquisitive mind.[3]

In vogue at that time was a gambling game which had been played for at least a hundred years, and which persists to the present day. It goes as follows: The "house" (that is to say, the professional gambler or the management of the gambling establishment)

[2] I am taking the position here that "birth" occurs when the infant really sees the light of day. Conception, of course, occurs earlier, and the gestation period may be a long one. In the case of Lady Luck it was long indeed. For Professor Oystein Ore of Yale has pointed out in his delightful book *Cardano, the Gambling Scholar* (Princeton University Press, 1953) that *Liber de Ludo Aleae* (*Book of Dice Games*), published in 1663 but written more than a century earlier, contained so much of the beginning of probability theory that "it would seem much more just to date the beginnings of probability theory from Cardano's treatise, rather than the customary reckoning from Pascal's discussions with his gambling friend de Méré, and the ensuing correspondence with Fermat."

[3] In this account, I am drawing heavily upon the historical researches of Professor Ore.

offers to bet even money that a player will throw at least one six in four throws of a single die.

In case this book falls into completely innocent hands (the probability of which seems small), I should perhaps state that a die is a small cube with slightly rounded corners and edges, presumably exactly symmetrical in shape and uniform in material (i.e., not "loaded"), whose faces bear the numbers 1, 2, 3, 4, 5, and 6 (see Fig. 6). When it is "thrown" or "rolled" or "tossed," a die should be as likely to

FIG. 6

come to rest on one side as on any other. And therefore it is reasonable to suppose that a die is as likely to "show," on its uppermost face, any one of the six numbers as any other one.

A couple of chapters further on we will see that the game just described is mildly favorable to the house. In fact, it turns out that the house, on the average, wins this game about 671 times to every 625 times it loses. But de Méré was not concerned with this game; he was puzzled by a similar but slightly more complicated one.

Suppose that instead of a single die, the player

rolls a pair of dice. De Méré was bothered about the question: Why isn't it favorable to the house to bet that the player will throw at least one *double six* (that is, a six showing on both dice of the pair) in *twenty-four* throws of the pair of dice?

Why did this question bother him? Almost surely because he had discovered, in this example, a conflict between a result he had himself calculated on theoretical grounds, and a somewhat different result which followed from a very old gamblers' rule.

This rule purported to tell — in a game of dice, say — what was the *critical number* of trials or throws critical number meaning the number of throws or trials at which the odds changed from *adverse* (for any lesser number) to *favorable* (for any greater number). The rule, when applied to this case, said that if *four* is the critical number of throws with a game with one die, as it in fact is, then *six times four*, or twenty-four, should be the critical number of throws in a game with two dice. The multiple of six arose, according to the rule, because two dice can come up in six times as many ways as one die can.

That two dice can come up in six times as many ways as one die can is clear from the fact that there may be associated with any given face on the first die any one of the six faces on the second. Thus, whereas for one die there are just the six cases:

CASES WITH ONE DIE

$$\left.\begin{matrix}1\\2\\3\\4\\5\\6\end{matrix}\right\}\text{six cases}$$

with two dice one has:

CASES WITH TWO DICE

Face of First Die	*Face of Second Die*	
1	1	
1	2	
1	3	
1	4	
1	5	
1	6	
2	1	
2	2	
2	3	
2	4	
2	5	
2	6	
3	1	
3	2	
3	3	
3	4	
3	5	
3	6	thirty-six cases
4	1	
4	2	
4	3	
4	4	
4	5	
4	6	
5	1	
5	2	
5	3	
5	4	
5	5	
5	6	
6	1	
6	2	
6	3	
6	4	
6	5	
6	6	

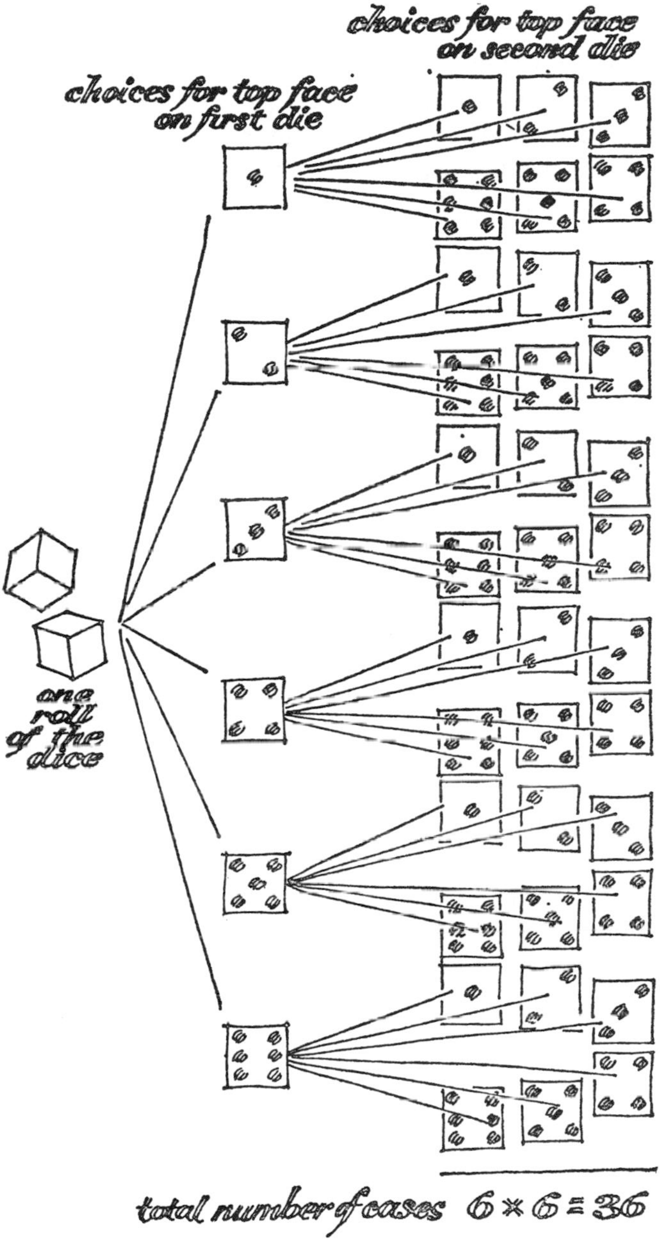
choices for top face
on second die
choices for top face
on first die
one
roll
of the
dice
total number of cases 6 * 6 = 36

FIG. 7

Rather than this rather clumsy and dull table, it is more illuminating to display by means of a different sort of diagram the total number of results, or cases, when two dice are thrown. The following branching tree diagram (Fig. 7) clearly shows that there are $6 \times 6 = 36$ cases, all different, for two dice. The combination 1 and 6 differs from the combination 6 and 1 because one is distinguishing between the two dice.

The old gambling rule thus indicated that the house could afford to bet even odds that any player would throw at least one double six in four times six, or twenty-four, throws of a pair of dice. But de Méré had apparently reasoned out that, in spite of the rule, twenty-four was not the right answer.

In many accounts of this story it has been assumed that de Méré actually discovered by sad experience that twenty-four was an unsafe number of throws to bet on. We will see, presently, that this is exceedingly unlikely; for the various margins of odds are so small that one would have to gamble on these games for a long time indeed to give a convincing experimental demonstration. Almost surely his objection was based on theory, not practice.

It is not known just how de Méré could have figured out the right solution. It is a very easy problem by present-day standards, and we will soon see how to solve it. But, in any event, de Méré consulted a young Frenchman named Blaise Pascal, whom he had met a short time before at a sort of house party arranged by another nobleman, to confirm whether or not he was right.

Pascal was a man of letters, a philosopher, a

theologian, and a mathematician.[4] He solved the problem put to him by de Méré, and proved that the odds were slightly *against* the house if it wagered on twenty-four throws, but were slightly *favorable* for twenty-five throws.

Pascal also solved a much harder and more significant problem also proposed to him by de Méré. This second problem was an old and famous one, which had been debated many times, but which had never been correctly solved. It was called the *problème des parties*, or as it is usually called in English, the *division problem*, or the *problem of points*. In general terms, the question is: How should the prize money be divided among the contestants if for some reason it proves necessary to call off a game, a contest, or a tournament before it is completed, and when the contestants have only "partial" scores?

Pascal introduced the very important idea that the amount of the prize any contestant deserved, in a partial game, should depend on the *probability* that this particular player would win the game, were it carried through to its conclusion. And Pascal worked out in detail, for several examples, how the probability of winning could be calculated from a knowledge of the nature of the game and the partial score of each contestant.

[4]He worked out for the first time the famous "arithmetic triangle," which exhibits the coefficients in the expansion of any binomial, and which we will describe in Chapter X. He eventually became a religious mystic, and in the latter part of his life doubtless shunned the gaming establishments, but he never lost his interest in probability questions. In one of his chief philosophical works there is a rather queer portion in which he is considering a *wager* as to whether or not God exists.

The questions of de Méré and the solutions furnished by Pascal proved to be very stimulating and significant.

Pascal wrote about these matters to another famous Frenchman, Pierre de Fermat, who had a great reputation as a mathematician and who was in addition a distinguished jurist at Toulouse. The resulting exchange of letters, which went further and further in working out the mathematics of some games of chance, became known in the learned society of their day. These results were, in fact, such an advance over any previous thinking about probability that this episode can properly be regarded as the real start of the mathematical theory. As Professor Ore has said,[5] "It was well known in Paris that Pascal and Fermat had discovered a *new branch* of mathematics." Within a short time the young Dutch genius Christian Huygens came to Paris to discuss these problems and solutions, and interest and activity in the new field grew rapidly.

Thus, in the highly fashionable but slightly disreputable atmosphere of the gambling rooms, and announced in the correspondence of French gentlemen of the seventeenth century, Lady Luck was born.

[5]But see the footnote on page 45, where Professor Ore expresses a somewhat different view.

Chapter III

THE CONCEPT OF MATHEMATICAL PROBABILITY

Probability is the very guide of life.
CICERO

The probable is what usually happens.
ARISTOTLE

Don't Expect Too Much

One of the fascinating things about probability theory is that it can be developed at almost any desired level of simplicity or of difficulty. There are some interesting remarks about probability that involve nothing more complicated than the ability to count on one's fingers. On the other hand, there is hardly any part of mathematics so deep, so general and abstract or so technically difficult and sophisticated, that it cannot be usefully applied to certain probability problems.

In this book we are not going to attempt the advanced topics; we will use only the tools of elementary mathematics. This restriction necessarily requires that the explanations will have to be just a little shaky at times, but we will try to be honest and give you some warning about the difficulties. You may be stimulated by the knowledge that there is a great deal more to learn about probability theory, and that some of it is so powerful and complicated that one must study mathematics courses, more and more difficult ones, for several years before being equipped to deal with the most advanced portions.

You need not feel apologetic or ashamed because there will be some slightly questionable episodes in our story, some slightly fuzzy or weak parts in our arguments. As is true of almost any aspect of life, you have to bring to this subject both courage and some faith. Rather than let yourself be paralyzed by minor difficulties, the thing to do is to go ahead and see what interesting and exciting and useful ideas can be developed.

Mathematical Theories and the Real World of Events

In Chapter I we saw that mathematical reasoning, based upon certain assumptions, develops the formal relationships which would be satisfied by any set of symbols or objects that "satisfy" the assumptions. Such a body of statements — assumptions and theorems derived from them — constitutes a pure mathematical theory. It may be elegant and beautiful, and it certainly is logically consistent; but it may or may not have any direct relationship with phenomena which actually occur in the real world.

If, for example, one of the assumptions is that A + B always equals B + A, then the theory will give useful results only for those symbols or real objects or acts for which order of addition is in fact immaterial. If the theory assumes (as ordinary plane geometry does) that the "points" with which it deals have no dimensions, but are simply pure, idealized locations, and similarly that the "lines" extend in one direction but have no thickness whatsoever, then one must expect that the theorems of the theory will hold with *complete* accuracy only for objects that correspond to those assumptions.

Of course, one cannot draw, with however sharp a real pencil, a line without width, nor make a dot of no dimension. But the theorems of geometry are nevertheless of very great practical usefulness. How does that come about?

It results, so sensibly as to be almost obvious, from the fact that real lines, real points, real triangles, etc., come closer and closer, the more carefully and accurately they are drawn, to the "ideal" lines, points, triangles, etc., with which the theory deals. And it results — again so sensibly that we often neglect to recognize how fortunate this all is — that the theorems of geometry do apply to *real* points and lines with a usefully high degree of approximation, and that we can force the degree of approximation to be better and better, if we do a better and better job of drawing the "real" figures. This illustrates the relationship between a "mathematical model" for a set of phenomena, and the actual relations exhibited by the phenomena themselves.

Mathematical Models

At various stages in the development of science one may consider a body of real phenomena and may try to invent a mathematical model — i.e., a set of assumptions plus the resulting body of pure theory which apply with strict accuracy to an idealized physical system which seems closely enough like the real physical system so that the theory of the idealized system will "explain," or at least organize and simplify the real phenomena. Or (if one is a very pure mathematician) one may be interested only in inventing and developing pure theories, leaving it to others to look around in the real world to see if there is any "application" for the theory; that is to say, whether there is a body of real phenomena behaving in accordance with the theory.

The reasons why one likes to have a mathematical model for a group of real phenomena are several and varied. The mathematical relations "express" the real relations in a highly condensed and useful way. The theories are often esthetically satisfying and intellectually stimulating. The mathematical relations are usually suitable for numerical calculation. With the mathematical model one can quickly and cheaply carry out the "paper" equivalent of real experiments, calculating what *would* happen under certain assumed conditions. It is, for example, a lot more economical to have a body of applicable theory concerning stresses and strains than it would be to build a bridge ten times stronger than it needs to be, or to start with a very weak one and then build stronger and stronger ones until finally one has built a bridge which sustains itself and its load.

Theoretically one could find the phenomena to which a theory applies by finding objects which exactly conform to the assumptions of the theory. But this, in general, is not a useful procedure. If we delayed using geometry until someone produced points with no dimensions and lines with no thickness, we would not be using geometry today.

Actually, the relation between the model and the real phenomena to which the results of the model are applied is a very complex one, difficult to analyze carefully. Starting with the facts of experimental experience and then becoming well and intimately acquainted with the way the real phenomena work, the scientist ordinarily arrives at a theory (i.e., a model) in a strange and wonderful way, compounded of experience, imagination, technical skill, art, and genius. He may, as did the early researchers, carry out some simple experiments which result in fairly crude measurements of quantitative relations; and then, bravely exerting a little imagination, he may idealize these measurements into a precise mathematical model. That was how Galileo enunciated the "laws" of a simple pendulum, these laws holding precisely true for a fictitious mathematical model, but holding with extremely useful accuracy for the small swings of real pendula. It was how Robert Hooke used different weights, and hence different forces, to stretch "elastic" materials, and then idealized his experience to produce the mathematical model in which the extension was precisely proportional to the force. Again, this law, holding strictly only for a fictitious mathematical model, has been extremely useful for predicting the actual elongation of bars of metal under tension, provided the forces and extensions are not too great. Every "law" that a civil engineer uses in designing a bridge, and presumably every law that an electrical engineer uses in designing a motor, is really a law which holds *precisely* only for a mathematical model, but which holds *approximately* and *usefully* for the actual material.

We are so familiar with, and have such justified

confidence in, a great many "scientific laws" that we tend to forget that very many[1] of them are really the theorems which hold strictly only for mathematical models, though they are obeyed closely enough by actual phenomena so that they are highly useful.

Can There Be Laws for Chance?

I have for an important reason stressed this matter of the relationship between real happenings and the theorems of mathematical models. In the case of many physical theories it is not too essential to insist on this relationship; one often has (for reasons that are in fact rather difficult to analyze) a satisfactory and comfortable sort of feeling that one "understands" the theory, and that it is unquestionably acceptable as the theory for the phenomena. If you pull on a rubber band it gets longer, and if you pull twice as hard it gets longer by twice as much. That seems perfectly reasonable,

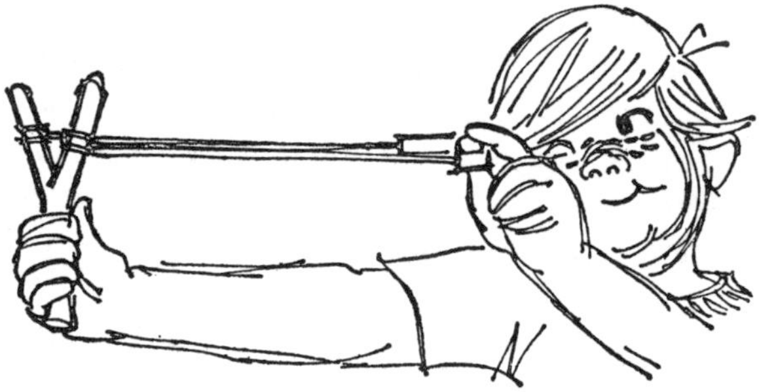

FIG. 8

[1]All of them? Think about this; it is a very interesting and deep point, which we must not interrupt our main job to consider.

even though it starts to get very complicated and even mysterious when you remember that the rubber band is made of molecules, and they of atoms, and they of elementary particles. But, on a large-scale view, you easily accept Hooke's law as being the theory which governs the extension of elastic substances under stress.

In the case of probability, however, the basic situation is quite different. When one tosses a coin, it may come up either heads or tails. There seems to be, at the level of simple common sense, no reason whatsoever for expecting that the continued tossing of a coin would exhibit numerical relationships (ratio of heads to tails, duration of runs of either heads or tails, large preponderance of heads over tails or tails over heads, etc.) which would satisfy "laws." A chance event is, one may say, a lawless event; so how can there be laws of chance?

Therefore in probability theory it is important to keep in the forefront of one's thinking this relationship between a mathematical model and the events of the real world. One does not start out, in probability theory, with a natural, intuitive, sort of understanding (perhaps I should say "feeling" rather than understanding) about what is happening. If someone tosses a coin a large number of times it seems sensible to suppose that about half the tosses will be heads and about half tails. But very likely it may seem to you equally sensible to suppose that it would be effectively *impossible* that, after a very large number of tosses, the actual number of heads would exceed the actual number of tails by, say, one thousand or even ten thousand. But (as we will see later) you would be wrong in this second hunch. And consider an event (such as your own death as the result of some operation) which, because of its nature, can occur only once. You die or you do not die — that is all there is to it. This certainly sounds, to recur to the phraseology of Chapter I, like an Aristotelian event for which it is proper to consider only two truth values, *yes* or *no*. How can one talk about the "probability" of an essentially non-repeatable event of this sort?

Mathematicians and philosophers and logicians have for a very long time argued about the "foundations" of probability theory. They have had very complicated difficulties in getting started in the theory, in a way which satisfied them. As we proceed, you will gradually understand, more and more clearly, why that was bound to be true.

But this confusion and these difficulties disappear[2] if one keeps in mind the "model — real world" concept. What one does, in probability theory, is invent a mathematical model, which can be calculated in a completely clear and tidy way, and then hope that this model will correspond in a useful way to some real phenomena. If, conversely, one faces a real problem, then through increasing experience one learns how to invent a model which is very likely to be useful for the real problem in question.

If all this sounds complicated and confusing, do not be discouraged. Plunge ahead, and then return to this section later.

The Rolling of a Pair of Dice

Let's start by going back to the simple business of rolling a pair of dice. To keep things straight, we will suppose one die is red, and the other is green.

[2]Or at least seem to fade — you just can't keep philosophers from arguing.

This makes it easy to tell them apart; and it turns out that it is important to distinguish between them.

Each cube is supposed to be a well-made, symmetrical, "fair" die. If you give one of them a good, vigorous roll, it is reasonable to suppose that the die is just as likely to land on one face as on any other. If you rolled it a very large number of times, it seems reasonable to suppose that very nearly one-sixth of the time it would show *one*, very nearly one-sixth of the time it would show *two*, and so on. That is to say, each face should get its fair share of appearances. On any one roll one face ought to be just as likely to appear as any other face.

Up to this point I have tried to describe the rolling of a real die in persuasive and "natural" terms. But (as you should guess from the preceding section) it is much more accurate and more rigorous to put it another way: Let us consider a mathematical model in which there is an idealized die which *by definition and assumption* is precisely as likely to show one of its faces, when rolled, as any other face. Let us go on talking about this idealized die of our mathematical model; and let us, for the moment, merely hope, or accept on faith, that there are real physical dice which come close enough to being like the idealized dice so that the resulting theory will apply to the real dice.

A moment ago we said about the fictitious die of the mathematical model that any one face is as "likely" to show as another. Actually, we will want to make this statement more specific in a quantitative way, but we postpone that for a bit.

Now, returning to our two dice, suppose we roll *both* the red die and the green one. Some of the time we will be talking about *real* dice, and some of the time about fictitious dice of our mathematical model.

We will try to keep clear which we mean, whenever the distinction is important.

Let's write down a list of all the different events or outcomes that could occur.

Case No. 1: The green die might show *one*, and the red die might show *one* also.

Case No. 2: The green die might show *one*, and the red die *two*.

Case No. 3: The green die . . .

We are obviously using more words than are necessary, so let's stop writing out a complete verbal description, and make a condensed table:

TABLE I

THINGS THAT CAN HAPPEN WHEN TWO DICE ARE ROLLED

Case Number	*Face Showing on Green Die*	*Face Showing on Red Die*
1	1	1
2	1	2
3	1	3
.	.	.
.	.	.
6	1	6
7	2	1
8	2	2
.	.	.
.	.	.
.	.	.
12	2	6
13	3	1
.	.	.
.	.	.
.	.	.
.	.	.
18	3	6

TABLE I (Continued)

Case Number	*Face Showing on Green Die*	*Face Showing on Red Die*
19	4	1
20	4	2
.	.	.
.	.	.
.	.	.
.	.	.
.	.	.
.	.	.
.	.	.
.	.	.
.	.	.
.	.	.
.	.	.
.	.	.
.	.	.
34	6	4
35	6	5
36	6	6

The dots, of course, represent all the cases we have not bothered to write down, but you can fill them all in if you have any doubts.

This table shows that with any one of the six faces on the green die there can, of course, appear any one of the six faces on the red die; so there are six times six, or thirty-six, outcomes[3] in all. We have written out only fourteen representative ones, and the dots, as we said before, stand for all the rest.

For the fictitious dice of our model we now simply *assign* equal probabilities to the thirty-six cases just displayed, so that each one of these cases will appear just about one thirty-sixth of the time.

[3]See the note on terminology at the end of this chapter which warns you that I am using rather old-fashioned no tation here.

Strictly, the resulting theory will, of course, be a theory of the fictitious model only. But we have in the back of our minds, all the while, that the model theory ought to work very well for real dice, provided they are "true" dice, symmetrical, homogeneous, and fairly rolled.

Suppose now that you are concerned about the *sum* of the numbers appearing on the two dice, as one is in many games with dice. How frequently can you expect to roll a sum of two, or of seven, or of eleven?

Among all the thirty-six possible outcomes there is only one, namely that listed first, which produces a sum of two. So in a large number of rolls you can expect to roll a sum of two, with a pair of dice, in just about one thirty-sixth of the rolls. But a sum of seven arises in six of the different possible outcomes. Namely:

TABLE II
CASES WITH TWO DICE WHICH PRODUCE A SUM OF SEVEN

Case No.	*Face Showing on Green Die*	*Face Showing on Red Die*	*Sum for the Two Dice*	
6	1	6	7	
11	2	5	7	6 different
16	3	4	7	outcomes, each
21	4	3	7	of which produces
26	5	2	7	a sum of 7
31	6	1	7	

And since each one of these six combinations happens in just about one thirty-sixth of a large number of rolls, seven comes up in just about six times one thirty-sixth, or six thirty-sixths, or just about one-sixth, of the rolls. These statements are surely correct for the fictitious dice of the model theory, and it is reasonable to hope that they are also correct for good real dice.

The similar table for a sum of eleven is:

TABLE III

CASES WITH TWO DICE WHICH PRODUCE A SUM OF ELEVEN

Case No.	*Green Die*	*Red Die*	*Sum*	
30	5	6	11	2 different outcomes, each of which produces a sum of 11
35	6	5	11	

So a sum of eleven should appear in just about two times one thirty-sixth of the rolls, or in one-eighteenth of the rolls.

In this way it is easy to figure out relatively how frequently, when one rolls a pair of dice, should appear sums of two, three, four, five . . . ten, eleven, and twelve. In the terms used by those who play the gambling game known as "craps," two (or "snake-eyes") and twelve (or "box-cars") are both "one-way points" very hard to make since they occur,[4] on the average, only once in thirty-six rolls. Three and eleven are both two-way points, which you can expect to roll about once in eighteen rolls on the average. Four and ten are three-way points, which should appear about once per twelve rolls in the long run. Five and nine are four-way points, and should appear about once per nine rolls. Six and eight are five-way points, which should appear just a little

[4]Here we have courageously jumped from the model to the real world!

better than once per seven rolls or, to state it more accurately, just about five times per thirty-six rolls. And seven, the critical number which wins in craps if you roll it first off, but loses if you roll it when you are trying to "make" some other point, is the only six-way point (see Table II), and should therefore come up about six times in thirty-six rolls, or once in six rolls, all these statements holding "on the average" or "in the long run."

You have every right to ask, at this juncture, "But *do* good real dice actually behave the way the model theory says they ought?"

The simple and plain answer to this is, Yes. A vast amount of exciting experience exists which bears on this question, as every crap-shooter knows. The theory *is* useful; it *does* work.[5] But it is only fair to remark that if you studied this subject a long time, you would then be prepared to face the fact that any precise analysis of the exact nature of the agreement between "theory" and "practice" involves some pretty detailed and fancy reasoning. You must not be surprised or disappointed at this, for the exact relation between theory and practice is complicated in any field.

When you make an experimental test of a the-

[5]A group at Duke University, under the leadership of Dr. J. B. Rhine, is convinced that the mind is capable of exerting a direct influence on physical events. In particular, they believe that they have demonstrated that an experimenter, sitting by a machine which rolls dice, and concentrating on the wish that the machine throw, say fives, can influence the machine to throw five (or any other number they concentrate on) significantly more than one-sixth of the time. Although it is difficult to refute their evidence, this result has not generally been accepted by scientists.

oretical probability statement, you do not expect to find an absolutely exact agreement. If one tossed a coin a million times and got *precisely* 500,000 heads, one certainly would be astonished; but one would be vastly more amazed if only 400,000 heads appeared.

In later portions of this book we will be ready to discuss somewhat better this relationship between theoretical and actual results in probability theory. For the moment we will say only that the model theory for dice is a useful one, because good real dice do behave just about the way the theory says they should.

The Number of Outcomes

One applies probability thinking to situations in which a number (sometimes a large number) of different outcomes can occur. What is needed here is some guide to the relative likelihood of these different outcomes.

All these alternative outcomes that might occur can, in general, be called simple *events*. If you roll one die, there are six possible simple events. If you roll a pair of dice, there are thirty-six possible simple events. If you toss a coin, there are only two simple events, head or tail, since tossed coins just do not land and remain on edge.

If you cut a deck of cards, how many simple events are there? There are 52, depending on which one of the 52 cards in the deck is exposed by the cut. But you may be interested only in the *color* of the card exposed. Then the *events* in which you are interested (note that I have left off the adjective "simple") are only two in number, namely "red" or "black," and each of these *events* is composed of 26 *simple events.*

If you are concerned with *suit,* then there are four *events,* each composed of 13 *simple events.* If you are concerned with the *rank* of the card, independent of its suit, then there are 13 *events* each composed of 4 *simple events.*

These words are not always used with precise meaning, but it is good practice, I believe, to use the words *outcome* or *simple event* for *all* the different things that can happen, and the word *event* for the simple event or set of simple events in which you are specially interested, and whose probability you wish to compute.

If you are considering a surgical operation, how many outcomes or simple events, and how many events, are there? There doubtless would be a wide range of possibilities, from complete and prompt recovery to death at the tragic other end of the list, with various degrees of recovery as intermediate

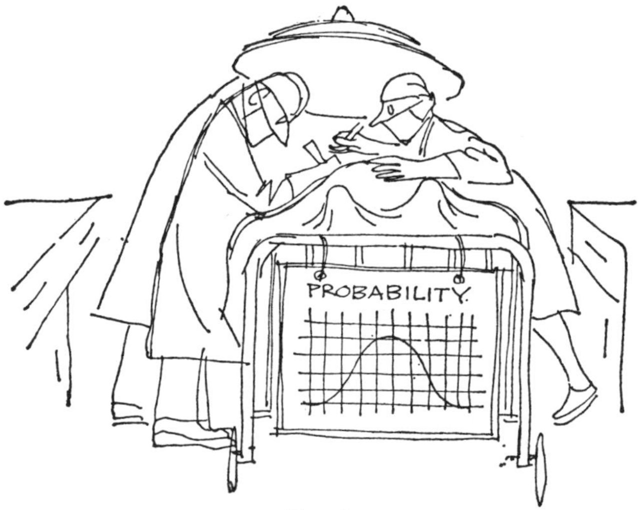

FIG. 9

possibilities. Thus there would be many outcomes or simple events. From one point of view, however, you might be concerned only with a single question: Do you survive the operation or do you not? That point of view would class all the possible outcomes into two kinds — those that lead to survival and those that do not.

If you are betting on a single ball game, then there are, for you, just three events for each team — win, lose, or tie. Each of these events might be composed of a lot of simple events if you take into account, say, the size of the score. If you are interested in a betting pool on the over-all outcome of all ten games played on a certain day in the American and National baseball Leagues, then there is a large number of outcomes, depending on all assortments of the possible individual outcomes for all ten games.

Equally Probable Outcomes

The various examples just given differ in an important regard.

When you fairly roll one well-made die, it seems reasonable to apply, to this real trial, the theory of the mathematical model which is based on the assumption that the six possible outcomes are *equally* probable. For the negative side,[6] you have no reason to suppose one face of the real die more

[6]It is interesting to note that it is not safe to assume various outcomes as equally probable on the basis of negative evidence, just because you lack any definite reason to favor one outcome over others. Once you begin conceding that various outcomes are equally probable, you find yourself caught in very tangled arguments which lead to fantastic conclusions. Plato used this sort of argument to "prove" the existence of Atlantis; and this sort of procedure can be used to "prove" that the moon is in fact made of green cheese.

likely to turn up than any other face. And on the positive side, this "true" die is accurately made of homogeneous material; and except for the very minor fact of different impressions on the various faces for the different numbers of dots, it is symmetrical.

For the same reasons, one feels comfortable and justified in applying the model theory to the rolling of two dice, and thus in treating the thirty-six possible outcomes as equally probable. On closely similar reasoning, it seems perfectly sensible to treat the fifty-two possible outcomes as equally probable when a well-shuffled deck of honest cards is given a fair cut. If you toss a symmetrical and "unloaded" coin, then again it is reasonable to consider that the outcome of a head is just as likely as that of a tail.

But in the other examples given in the preceding section, it is obvious that the various outcomes are not equally probable. If, for example, the surgical operation is a very routine and minor one, and if you are in vigorous health, then it is extremely likely that you will survive the operation, and extremely unlikely that it will prove fatal. This case has a further important peculiarity. A large part of one's willingness to consider as equally probable the various outcomes with dice and cards and coins depends on the fact that one can roll and cut and toss as many times as one pleases, and can observe that the "equally probable outcomes" do in fact tend to happen equally often. In the case of the operation there is, of course, the record of surgical experience in *similar* cases. But the operation on *you*, with all your special qualifications, can be made only once.

In the case of the ball game the likelihood of a win, or of a loss, or of a tie, would normally differ, and the exact differences would depend on many factors of skill and luck.

Ways of Designing Models

One finds, in general, two kinds of situations in studying real phenomena which involve probability considerations. In one situation the investigator, after thoughtful examination, can say, "There are involved here thus-and-so many possible outcomes, and it seems perfectly sensible to consider them all equally probable."

This first type arises chiefly in games (as we have just seen), and one has no difficulty in setting up a model, which can be expected to give results with useful application to the real situation. The model is, in fact, simply the idealized game, played with perfect dice perfectly rolled, or with perfect coins perfectly tossed, or with perfect decks of cards perfectly shuffled and cut or dealt. In each case the connotation of the word *perfect* is "so that the various possible elementary outcomes are equally probable."

Or — and this is the second type of situation — it may just not be possible to sort out a set of outcomes of such a nature that it seems sensible to assume them to be equally probable. A great many of the practical questions of life, about taking an umbrella, or deciding to try a new vitamin mixture, or the more serious problems of running risks in business or in war, fall in this second category of situations, in which it is not possible, or at least not sensible and convincing, to try to designate a lot of equally probable outcomes.

But here again we have to have a model in order to have a model theory; and the problem is, how do we design the model so that its theory can reasonably be expected to apply usefully to the real situation?

I will give two illustrations of this second type of situation. Suppose you are interested in the voting of a large community composed of some Democrats, some Republicans, some Socialists, and some Communists. Suppose the idea is to select a "representative" sample; and suppose you are wondering how likely it is that a sample of, say, 200 chosen by some sensible scheme will represent fairly (to such-and-such accuracy) the actual proportions of the various party members in the community. Or, if you decide that you want to be 98 per cent sure of getting a sample that will give the proportions with accuracies of at least 95 per cent, how large must that sample be?

The early experts in probability theory were forever talking about drawing colored balls out of "urns." This was not because people are really interested in jars or boxes full of a mixed-up lot of colored balls, but because those urns full of balls could often be designed so that they served as useful and illuminating models of important real situations. In fact, the urns and balls are not themselves supposed real. They are fictitious and idealized urns and balls, so that the probability of drawing out any one ball is just the same as for any other. So far, that sounds like the former case of equal probabilities for all outcomes.

But in an urn of 1000 balls you can put 300 which are white, 500 which are red, and 200 which are black. And by varying these numbers, you obviously can produce a model in which you can arrange, just as you may please, the relative probabilities of drawing a white, a red, or a black ball. If in an urn you put a large number of balls, a proper proportion being white (Democrats), a proper proportion blue (Republicans), a proper proportion green (Socialists), and a proper proportion red (Communists), then you can work out the pure theory of this model (in a way we will learn presently), and obtain convincing answers to the original question about the real voters.

Or suppose some anxious parents worry about the possibility that one or more of their children contract, within a period of one year, say, a (non-communicable) disease present in their community. Neglecting all possible influence of hereditary type, diet, family circumstances, etc., the parents might note that in recent years about four out of every 100,000 children in the relevant age range contracted the disease in their community.[7]

One can then think of a box with 100,000 balls, 99,996 of which are white and 4 black. What is the probability of drawing, in succession, three black balls? Would drawing balls out of this box usefully simulate the real situation of contracting, or not contracting, the disease? In this case one has a good many aspects to consider before deciding that the model is enough "like" the real case so that in practice the theoretical calculations concerning the model mean very much.

It would be easy to go on with other illustrations in which it is still harder to invent a model of

[7]These are sensible figures for polio, at the present time, in the absence of any epidemic.

convincing usefulness. But the point you must hold on to is this: probability calculations (except for essentially trivial ones) apply strictly only to the fictitious models on which they are based. Whether or not they apply usefully to the real situations which were in your mind when you invented the models is a point you ought always think carefully about. Probability theory is wonderful, and wonderfully helpful, but it isn't magic.

The Definition of Mathematical Probability

Let us agree that we are, for the moment, thinking about an experiment or a trial which can result in N *different equally probable outcomes.* You understand that we are talking about a *general* situation, not a *specific* one; so we do not at the moment know (or care) what N is. It may be 2, or 6, or 36, or 52, or any other positive integer. Let us further agree, as a matter of general notation, that of all these N equally probable outcomes, we consider n of them (again n is a positive integer, general except for the condition that it clearly cannot exceed N) as "favorable" in the sense that the occurrence of any one of them results in the event we are interested in, or are concerned about, or desire. Let us call this event E.

We are now ready to state the definition of the mathematical probability of an event: *The probability $P(E)$ of an event E is defined by the equation:*

$$P(E) = \frac{n}{N} \tag{1}$$

where N is the total number of equally probable outcomes, and n is the number of the outcomes which *constitute the event E*. The word "constitute" in this definition may be used in a rather queer way. The n outcomes *constitute the event E*, in the sense that if any one of them occurs, then E is said to have occurred.

It has been for the moment assumed that both n and N are integers—whole numbers, but this definition can be generalized to cover much broader possibilities.

If the event E occurs in *all* the N outcomes which can happen — that is to say, if every outcome results in E — then $n = N$ and $P(E) = 1$. But if all possible outcomes result in E, then E is assured. A mathematical probability equal to 1 corresponds to the *certain* occurrence of the event E under consideration, or more briefly, and more generally, to *certainty*. On the other hand, if none of the N ways results in E, then $n = 0$ and $P(E) = 0$. That is to say, a mathematical probability of 0 corresponds to the *impossibility of occurrence of E*, or more briefly, to the *impossibility* of E.

The older books on probability theory often used the terminology "the probability of success," (meaning the same thing as "the probability that the event E occurs), and the expression "the probability of failure" (meaning the same thing as "the probability that E does not occur). But the event E may well be the death of a certain child, or nuclear war, so it seems more sensible to use the neutral expression "the event E occurs" rather than the loaded words "success" or "failure."

In many games it is customary to speak of the *odds* for (or against) an event E, rather than the *probability* of the event E. The relation between the two terminologies is a simple one. When the probability is n/N, there are n outcomes which result in E, and $(N - n)$ which do not result in E. Thus we can make two statements:

$$\begin{cases} P(E) = \dfrac{n}{N} \\ \text{The odds in favor of } E \text{ are } n \text{ to } N - n \end{cases}$$

These statements are exactly equivalent.

The equation $P(E) = 1/2 = 0.5$ says that n is exactly one-half of N, or that just one-half of the possible outcomes result in E. The statements "the probability is 0.5," "the odds are one-to-one," "the odds are even" are exactly equivalent statements.

If you are throwing a die and want an ace to show, then $n = 1$, $N = 6$, and $(N - n)$ is 5. Thus the *probability* of any one face's showing is 1/6, whereas the *odds* in favor of an ace, say, are 1 to 5. The odds against throwing an ace are 5 to 1.

It is critically important to realize that this definition of probability (in Equation 1) is of no use at all unless applied to situations in which the phrase "equally probable outcomes" is applicable. There is no point in concealing the fact that this last remark is a very important one, which, if you are philosophically inclined or very rigorous about logical precision, presents a formidable difficulty. There is no difficulty about "equally probable" in the case of the mathematical models, for one simply *assumes* this characteristic of the model. But when one raises the inevitable question, "Do the model results apply to the real case?" he is bound to think about the "equal probability" of real occurrences. Just how can one really judge that actual outcomes are equally probable? Even more fundamentally, what precisely does one mean by saying that real outcomes are equally probable?

You should not be discouraged about these complications nor should you use this as an excuse for terminating your interest in probability, any more than you would discard geometry because of the logical gap between geometrical theory and its practical applications in mechanical drawing or land measurement. This is the way science and knowledge progress — by uncovering difficulties, by clearing up some of them, by accepting the challenge of those that remain.

And as I have implied in previous remarks, be encouraged and stimulated by the indisputable fact that probability theory *works*, in the sense of practical usefulness and success. You probably cannot give a logically satisfying description of the fact that you are able to walk — but you do not permit this to immobilize you!

A Recapitulation and a Look Ahead

Before we proceed, let me give you a little résumé of what has happened so far, and of how things are going to work out in the next couple of chapters.

After a preliminary chapter about thinking and reasoning, and a historical chapter which tells how the theory of probability got started three centuries ago, we have given, in this third chapter, a definition of the mathematical probability of an event.

This definition can, strictly speaking, be used only in rather limited and rather artificial circumstances

— namely, those in which it is possible to enumerate the total number N of outcomes which *could* occur in a trial or operation, and *only* when these N outcomes are all reasonably to be judged as equally probable. Strictly speaking, this system works only in situations such as those found in the idealized mathematical models which one invents, hoping they will simulate real situations.

That doesn't sound too promising. It is exciting to play games, and amusing to draw balls out of jars, but how about *real* problems?

Let me, very roughly at this stage, give you a forecast of what is going to happen in the succeeding chapters.

We are going to find that if the probability of an event (reckoned in accordance with the definition just given) is, say, 0.57, then if one conducts a long series of trials, the proportion of successes for the event in question tends to differ less and less from 0.57 (or the percentage of successes, if you prefer that terminology, less and less from 57 per cent). This fact is proved in a theorem which is the most important and central fact of probability theory. We will get to this theorem, and discuss it, in Chapter XI.

The result just stated furnishes a major bridge (even though experts in logic recognize it as a somewhat shaky bridge) between the artificial *mathematical* theory of probability, based on the definition, which restricts one to rather artificial cases, and the useful application of that theory to all sorts of real and practical situations.

Even at this early stage we can get a glimpse of how this works.

Let us return to the worried parents we mentioned a little time ago, and examine their problem a little more closely. Suppose they, as the parents of several children, are concerned about the probability that one of their children will get paralytic polio; and also about the even more serious possibility that two or more of them will get this awful disease. Quite apart from all the more human aspects, they might have money enough to survive the costs of *one* such illness in the family, but not *two*.

As you know, the health authorities keep records of the incidence of diseases. Suppose that the incidence of paralytic polio (for the right age group, the locality in which you live, etc.) is four in one hundred thousand. Well, *if* you were doing a very long series of trials of drawing balls, four out of every one hundred thousand being black balls, the fraction of black balls you would draw would tend to get closer and closer to four black balls per every hundred thousand balls drawn out.

Thus this box of white and black balls may serve as a useful model for answering questions about the incidence of paralytic polio.

This model for example would say, "If a ball is drawn at random, there are four chances in one hundred thousand, or one chance in twenty-five thousand, of drawing a black ball." This seems reasonably enough like the "real" polio situation so that you are prepared to say: With the stated incidence, the chance that any one child will get the disease is one in twenty-five thousand, or the probability is 1/25,000.

We have crossed the shaky bridge! For we have made a statement about the mathematical probability that a child will get polio, in spite of the fact that we cannot describe or exhibit "N equally likely cases."

If one draws out two balls, what are the chances that they will both be black (i.e., that two specified children both will get polio)? If one draws out five balls, what are the chances that at least one of them will be black (i.e., the chance that at least one of the five children will get polio)? What are the chances in drawing five balls that at least two will be black? Etc., etc.

All kinds of questions of this sort can be asked about the mathematical probability of drawing white and black balls out of this box. I am sure that you can see that these questions and answers might bear a most illuminating relationship to vitally important questions about the chances that one or more children in a family will get a dread disease.[8]

As we proceed to the following chapters, all this should become more definite, and more clear.

Note on Terminology

This book is written to interest you in probability. If it also persuades you to go further into the subject, you will discover that modern writers of textbooks and of research papers use terminology you will not find here. Realizing that most of my readers will find a good many of the ideas new (and hence strange), and some of the reasoning a bit complicated, I have thought it sensible to use the minimum

[8]If the disease is communicable, or if it is under the influence of genetic or environmental factors, then the idealized model would have to take all that into account.

number of new and strange words and phrases, even though that restriction forces me to use *more* words than would otherwise be necessary.

In modern parlance the collection of the 36 possible outcomes which can occur when a pair of dice is thrown is called a *sample space* generated by throwing a pair of dice. This terminology is a very useful one to a person who goes on to study the theory seriously. A sample space of an experiment can be given a neat and precise definition, and from then on the term serves compactly to express precisely what otherwise requires more words and may achieve less precision. But until they have worked a fair bit with the phrase and the idea, those who are not mathematically minded may find the term "sample space" queer and troublesome. I can't avoid making you worry a bit, and I certainly want to stimulate you to think; but for this "first time over lightly" I don't want you to trip up on unfamiliar words and phrases unless it is really essential.

Note on Other Books about Probability

There are many, many interesting and good books about probability theory. I will list here five books, which seem to me excellent, the list starting with two items rather more elementary than the present book. The other three are essentially textbooks of increasing scope and depth of treatment.

1. The Science Materials Center at 59 Fourth Avenue, New York 3, N.Y., has a Probability and Statistics Kit, accompanied by a fine little instruction booklet by Edmund C. Berkeley. The kit contains

materials to carry out a number of probability experiments.

2. *The Science of Chance*, by Horace C. Levinson (New York: Rinehart & Company, 1950), contains a very interesting and authentic description of probability, statistics, and applications, involving no equations and a minimum of mathematical reasoning.

3. *An Introduction to Finite Mathematics*, by Kemeny, Snell, and Thompson (Englewood Cliffs, New Jersey: Prentice-Hall, 1957), is a very lively text for an introductory college course in mathematics. It contains good material on probability.

4. *Probability and Statistics*, by Mosteller, Rourke, and Thomas (Reading, Massachusetts: Addison-Wesley Publishing Co., 1961), is the official text which was used for the Continental Classroom TV lectures on probability which Professor Mosteller gave so brilliantly and effectively. It is a mathematical text, but is extraordinarily clear and interesting.

5. *An Introduction to Probability Theory and Its Applications*, Vol. 1, Second Edition, by William Feller (New York: John Wiley & Sons, 1957), is a much more advanced text by one of the leading experts in mathematical probability.

Chapter IV

THE COUNTING OF CASES

The theory in question [probability] affords an excellent illustration of the application of the theory of permutations and combinations.

GEORGE CHRYSTAL, *Algebra*

Preliminary

The numerical value of the mathematical probability of an event E being defined as the ratio of the number n of outcomes which result in E to the total number N of equally probable outcomes, it is of obvious importance to be able to count, or otherwise enumerate, the number n and the number N for the problem at hand.

In the almost trivially simple examples of tossing a *single* coin or one or two dice, or in cutting a deck of cards, the problem of counting is easy indeed. But if you think of even so simple and familiar a matter as throwing five coins, or rolling six dice or dealing poker or bridge hands, the counting of the various outcomes which constitute an event of interest is neither obvious nor easy.

A considerable range of problems can be handled by methods learned in elementary algebra. We will take just a little space and time to recall some of these results; then at the end of this chapter I will have something to say about certain more modern and more powerful methods.

Compound Events

A compound event is one which, as the adjective suggests, is composed of two or more component events.[1] Thus a roll of two dice is a compound event, composed of two separate component events; namely, the rolling of one die, and the rolling of the other die.

Suppose we think first of a twofold event, composed of only two component events, and suppose that this twofold event consists in the simultaneous tossing of a coin and the rolling of a die. How many different outcomes result? As to the coin, it can come up in two ways — heads or tails. And whichever the coin does, the die can quite independently come up in any one of six ways. Thus there are six outcomes of the die which may accompany heads on the coin, and six outcomes of the die which may accompany tails on the coin. Thus there are *two* times *six* or twelve outcomes of the compound event in question.

If a twofold event consists in rolling two dice, then with each of the six faces on the first die there can be associated any one of the six faces on the second. Thus (as we saw in the preceding chapter) there are *six* times *six* or thirty-six outcomes for this particular compound event.

[1]In the technical modern terminology referred to in the note at the end of the last chapter, one would say, "A product sample space is the product of two or more sample spaces."

If we consider a triply compound event, such as the simultaneous tossing of a coin and the rolling of two dice, the total number of outcomes would be $2 \times 6 \times 6 = 72$.

Thus it is easy to see what the general rule is. If the compound event in question is composed of several component events, one of which has a outcomes, a second of which has b outcomes, a third of which has c outcomes, etc., then the total component event has a number of outcomes which is the product[2] of a by b by c, etc.

Permutations

If you have a number of different objects (which may be simply numbers or letters), in how many ways can you put them in lines, each line to contain all the objects, and the *order* of the objects to differ from line to line? In shorter terms, in how many ways can you *order* them?

Suppose we have the first three letters of the alphabet, A, B, and C. In how many different orders can I write them down? ABC is one such order, and BCA is another, and CBA is another. How many are there in all?

Suppose I have three boxes in a row. I must put a letter (each time a *different* letter) in each box. This is a triply compound event. I can fill the first box in three ways, for all three letters are available for

[2]You can see now that calling the compound event "a product sample space composed of the product of the individual sample spaces" has the verbal advantage that the word "product" leads naturally to the statement to which this footnote refers.

the first choice. I can then fill the second box in two ways, since there are now only two letters left to choose between. And I can fill the third box in only one way, because by that time I am stuck with whatever letter is left over.

Thus by the principle for compound events, and as one also can easily see by drawing a branched diagram similar to Fig. 7, I can fill the three boxes in $3 \times 2 \times 1 = 6$ ways.

The general rule is now obvious. If I have n objects, all n are available for choice for position number one, so I can fill this position in n ways. Then $n - 1$ are left as candidates for position number 2, which I can thus fill in $n - 1$ ways. I can fill the third position in $n - 2$ ways: and so on. Clearly if ${}_nP_n$ denotes[3] the total number of orders of n objects —that is, the number of *permutations* of n objects (for that is what is meant by a "permutation"), then

$$ {}_nP_n = n(n-1)(n-2)(n-3) \cdots 3 \times 2 \times 1 \qquad (2) $$

The right-hand side of this equation says that you must multiply n, the number of objects being permuted, by each successively smaller integer until you have finally run down to unity, this corresponding to the fact that there is only *one* way to fill the final box.

If you have n objects, but want to make ordered arrays each consisting of r objects (r being less than n), then ${}_nP_r$ or "the number of permutations of n

[3]In the next paragraph you will see why there is a subscript n written both before and after the P.

things taken r at a time," as the technical phrase goes, is clearly

$$ {}_nP_r = \underbrace{n(n-1)(n-2)\cdots(n-r+1)}_{r \text{ factors}} \tag{3} $$

You end up with the r'th box which can be filled with any of $(n - r + 1)$ things, that being the number left after filling the first $(r - 1)$ boxes.

The product

$$ n(n-1)(n-2)\cdots 3 \times 2 \times 1 $$

of all the integers from n down to 1 is a quantity appearing so often in the algebra of enumeration that it is convenient to give it a name, and abbreviate it to a single symbol. It is accordingly called *factorial n* or sometimes *n factorial.* It is usually written as $n!$, and sometimes as $\underline{|n}$, the latter being now viewed as rather old-fashioned.

In terms of this neat notation, the two previous equations read

$$ {}_nP_n = n! \tag{4} $$

$$ {}_nP_r = \frac{n!}{(n-r)!} \tag{5} $$

You can easily satisfy yourself that the $(n - r)!$ in the lower part of the last written fraction just kills off, by cancellation, the unwanted tail of $n!$ so as to make it properly stop with the r'th factor $(n - r + 1)$.

It is interesting, and important, to notice that this quantity *n factorial* grows very rapidly as n increases. The first few values are

TABLE IV
FACTORIALS

1!	=	1
2!	=	2
3!	=	6
4!	=	24
5!	=	120
6!	=	720
7!	=	5040
8!	=	40,320
9!	=	362,880
10!	=	3,628,800

The total number 52! of all possible orders for the 52 cards in a deck of 52 ordinary cards is a number of 68 digits which, if printed out in full, would require more than one full line of type on this page. If every human being on earth counted a million of these arrangements per second for twenty-four hours a day for lifetimes of eighty years each, they would have made only a negligible start in the job of counting all these arrangements — not a billionth of a billionth of one per cent of them! You can see that

FIG. 10

to handle even the very mild case of a deck of 52 cards it is necessary to have some mathematical tools with which to work.

We want to say a word, for those interested, about this curious and strange quantity *factorial n*, which increases so prodigiously as n increases. But first, for those who may have forgotten almost everything they were exposed to in high school algebra, we must say a word about *exponents*.

Frequently one has to multiply a number by itself, a certain number of times. Instead of writing

$$\underbrace{10 \times 10 \times 10 \times 10 \times 10 \times 10}_{\text{6 identical factors}}$$

one writes

$$10^6$$

The exponent 6 means precisely, "Multiply the number I am attached to by itself until there have been, in all, 6 factors."

Thus 10^6 is a short way of writing 1,000,000 or one million; and 10^{50} is an obviously convenient way of representing 10 multiplied by itself until there are 50 factors — that is to say, 1 followed by 50 zeroes.

Since

$$\underbrace{10 \times 10 \times 10}_{\text{3 factors}} \text{ multiplied by } \underbrace{10 \times 10 \times 10 \times 10 \times 10}_{\text{5 factors}}$$

is obviously

$$\underbrace{10 \times 10 \times 10 \times 10 \times 10 \times 10 \times 10 \times 10}_{3 + 5 = \text{8 factors}}$$

it follows that $10^3 \times 10^5 = 10^{3+5} = 10^8$

More generally, for any "base" number x,

$$x^a \times x^b = x^{a+b}$$

so that to *multiply* one *adds exponents*, the base being the same (x) in the two cases.

Similarly, canceling out three common factors in the numerator and the denominator, we have

$$\frac{\overbrace{10 \times 10 \times 10 \times 10 \times 10 \times 10 \times 10}^{7\text{ factors}}}{\underbrace{10 \times 10 \times 10}_{3\text{ factors}}} = \underbrace{10 \times 10 \times 10 \times 10}_{7-3\text{ factors}}$$

and so the general rule

$$\frac{x^a}{x^b} = x^{a-b} = x^a \times x^{-b}$$

This is all very simple, but very neat and useful.

The quantity $n!$ is a rather intractable one from the mathematical point of view. There is, for example, no easy way to multiply two factorials. So it is hard to manipulate equations containing this quantity. Fortunately, this difficulty can be avoided. For by methods learned in calculus it can be shown that $n!$ can, when n is good-sized, be approximated quite closely by the quantity

$$n^n \times e^{-n}\sqrt{2\pi n} \qquad \text{(A)}$$

where $e = 2.71828 \cdots$ and π has the familiar value $3.14159 \cdots$

You recognize, in the expression (A), that n^n means n multiplied by itself until there are n factors; whereas e^{-n} means "multiply $1/e$ by itself until there are n factors."

It is curious that the approximation (A), which *looks* far fancier than does $n!$, is ever so much handier than the simple-looking $n!$ For the pieces of the approximate expression are easy to calculate, and relatively convenient to manipulate in theoretical arguments.

The approximation (A) is a very good one in one sense, and a bad one in another sense. Since the point involved is a basic one that will come up later, we had better be clear about this matter.

There are two ways of comparing two numbers. You may be concerned, in an absolute sense, with the size of the two numbers, and thus with the actual *difference* between their sizes.

Or you may be concerned, in a relative sense, with the *ratio* of the two numbers. How far is this ratio from unity, or (to put the matter in other terms) by what *per cent* is the greater one of the two larger than the other?

These two viewpoints lead to two very different descriptions. Two numbers can differ by a large amount in the absolute sense, but differ very little in the ratio sense.

All this is well illustrated by the nature of the approximation to factorials which is furnished by expression (A). Look at this table:

Number n in question	*Factorial n*	*Expression (A)*	*Absolute Error in Expression (A): or Difference*	*Relative Error in Expression (A): or Ratio*	*Per Cent Error*
1	1	0.9221	.0779	.9221	7.8%
2	2	1.919	.081	.959	4.1%
5	120	118.019	1.981	.983	1.7%
10	3,628,800	3,598,600	30,200	.992	0.8%

You see that as n increases the absolute error in expression (A) increases. Indeed, it increases rapidly and becomes very large, as is illustrated by the fact that the absolute error is less than two when $n = 5$, but is over 30,000 when $n = 10$.

On the other hand, the relative error, the ratio, keeps getting better. For $n = 1$ the per cent error is nearly 8 per cent. For $n = 2$ this is roughly halved. For $n = 10$ the per cent error is less than 1 per cent. By the time n is as large as 100, the absolute error is exceedingly large, but the relative error for $n = 100$ is only 0.08 per cent.

Expression (A) is known as the Stirling Formula for *factorial n*. A Scottish mathematician named James Stirling worked it out more than two centuries ago.

Combinations

Sometimes one is not concerned about the *order* of the objects (letters, cards, numbers, or whatever), but is only interested in the *constitution* of the sample in question. Thus suppose we are dealing with the five letters A, B, C, D, and E. How many essentially different groups or *combinations*, to use the technical term, of three letters can one make out of these five letters? ABC is one combination. ABD is another, for you call it a different combination if only one element differs. So ABE is another, BCD another, and so on. How many in all are there?

With only a few items one could get the answer just by writing them all down in some systematic way. But it is important to have a general formula which applies no matter how many items there are.

If you ask: How many combinations can I make of five things if I take them *all?* Clearly there is only *one* such choice. I just take all the five items available — and that is that. Indeed, the number of combinations of n things, taken all n at a time, is clearly *one*, no matter what n is. In the case of the *permutations* of n things, it is important to talk about the number taken n at a time, or taken r at a time, r being less than n. In the case of *combinations* of n things, it is trivial to take all n, so one proceeds at once to the general formula for the "combinations of n things taken r at a time."

This general formula can be derived by a very simple trick. Suppose we denote *the number of combinations of n things taken r at a time* by $\binom{n}{r}$. In older books you may see the notation ${}_nC_r$ which is naturally related to the notation for the permutations of n things taken r at a time, but nowadays this older notation for combinations is seldom used.

Now suppose we take each one of these *combinations*, which contains r elements, and rearrange (or "permute") these r elements so as to obtain all the permutations or arrangements of these particular r elements. We could, as the previous section proved, produce $r!$ arrangements of these r elements.

Now do this with another of the combinations, and with another, and so on — until you have made $r!$ different permutations out of each one of these $\binom{n}{r}$ number of combinations. Every single one of the r-fold permutations thus made will differ from every other one, either in *constitution* (because of having started out with a different combination) or in *order* (because of being different permutations of that combination).

The totality of all of these r-fold permutations will obviously be all the r-fold permutations one can make, starting out with n objects. But that total number is known, from the previous section, to be ${}_nP_r$ as given by formula (5).

Thus we have $\binom{n}{r}$ combinations, each one of which makes $r!$ permutations, and the totality is ${}_nP_r$. In other words

$$r! \cdot \binom{n}{r} = {}_nP_r = \frac{n!}{(n-r)!}$$

or

$$\binom{n}{r} = \frac{n!}{r!\,(n-r)!} \tag{6}$$

I hope that you are not bored by this argument. You ought, in fact, to be thrilled! For here is a formula which is valid when n and r are mild little numbers such as three or four or five, but which is equally valid if n and r are in the thousands, millions, or billions, or more. This is a formula, derived by simple but rigorous and general thinking, that applies just as well to the number of states of the astronomical numbers of molecules in a gram of matter as it does to the six faces of a die.

Let us look at just a couple of illustrations of the use of the formulas we have derived.

An English word depends on the letters used, and on the order of the letters. How many four-letter words (which seem so popular of late) can one make out of the twenty-six letters in our alphabet? If we neglect the fact that it is customary to have at least one vowel in a word (but not absolutely essential, as the useful word "pfft" shows), and also the fact that letters can be repeated, even in so short a word, then the maximum number of four-letter words is

$$_{26}P_4 = 26 \cdot 25 \cdot 24 \cdot 23 = 358{,}800$$

which seems to set some sort of numerical upper limit to four-letter nastiness of the English language!

Or, to re-establish the moral tone of the book, if there are eighteen little girls of about the same age in a Sunday school, and three teachers available for them so that the classes will each contain six little girls, how many different choices confront the teacher who is permitted to select her class first? Here order is not important, as it is when spelling words with letters. Therefore we want to know how many combinations are there of eighteen "things," taken six at a time.

The answer is

$$\binom{18}{6} = \frac{18!}{6!\,12!} = \frac{18 \cdot 17 \cdot 16 \cdot 15 \cdot 14 \cdot 13}{6 \cdot 5 \cdot 4 \cdot 3 \cdot 2 \cdot 1} = 18{,}564$$

More Complicated Cases

Starting with the basic permutations and combination of n objects, there is an almost endless array of more complicated cases, some of which require very involved and tricky reasoning. We will not be using these fancy formulas, but you ought to look at a few, just to appreciate how powerful such methods can be.

Up to this point we have been dealing with objects which are all different. But suppose (as though you were a typesetter with a certain number of "m's," a certain number of "P's," etc., available in your supply of letters) you ask, What is the total number of permutations of n objects, of which a are alike and of one sort, b of which are alike and of another sort, etc.? The answer is

$$\frac{n!}{a!\,b!\cdots}$$

(We will outline the reasoning which establishes this formula in a problem at the end of Chapter VI).

Or suppose you have n distinct objects. In how many ways can you choose from them a group or combination of α objects, a second combination of β objects, a third of γ and so on? The answer is[4]

$$\frac{n!}{\alpha!\,\beta!\,\gamma!\cdots(n-\alpha-\beta-\gamma-\cdots)!}$$

Or, in how many ways can P identical objects be placed in n distinct boxes? The answer to this one (which arises in various situations in theoretical physics) is

$$\frac{(n+P-1)!}{(n-1)!\,P!}$$

Or suppose you ask: In how many ways can n beads of different colors be strung on a circular string? Remember that a string of beads can be turned over, and still be the same string, and that there is no uniquely defined starting point for the beads on a circular string. Thus with seven different beads of different sizes 360 different necklaces can be made; with six beads of the same size and one larger one, only one can be made; with beads of two sizes,

[4]Don't let these Greek letters frighten you. Mathematicians love to use them.

five of the smaller and two of the larger, three necklaces can be made.[5]

As the English mathematician Peter Nicholson stated nearly a century and a half ago,[6] "Combinatorial analysis is a branch of mathematics which teaches us to ascertain and exhibit all the possible ways in which a given number of things may be associated and mixed together, so that we may be certain that we have not missed any collection or arrangement." This sort of analysis has, from the very birth of Lady Luck, been essential to the development of probability theory.

Many textbooks on advanced high school algebra give examples of the formulas and problems. There is a large collection in a book called *Choice and Chance* by Whitworth, and in Chrystal's classical two-volume algebra.[7] Textbooks on probability theory give complicated (but useful and important) examples, and the "problem" sections of mathematical journals give curious and interesting examples almost every month. Just after I wrote this sentence I took off my shelves, at random, a recent copy of the *American Mathematical Monthly*. It proved to be the November 1960 number and, sure enough, it had a short article with the title "Combinatorial Miscellanea." And in the problem section there was this:

[5]This problem is taken from the book (see page 82) by Kemeny, Snell, and Thompson (p. 96). It gives a first-rate introduction to probability theory.

[6]See P. A. MacMahon's *Combinatory Analysis*, a two-volume work of 690 pages, entirely devoted to this special sort of problem.

[7]*Textbook of Algebra* (paperback, Dover).

"In the game of bridge North is dealer. (a) What is the number of the different bidding sequences, assuming that West and East are silent? (b) What is the number of the bidding sequences if all four players must participate in the bidding?"

The answer to (a) is the tremendously large number $(2^{36} - 1)$ and to (b) is the vastly larger number $(4 \cdot 22^{35} - 1)/3$.

In other words, the enumerating of cases by these more or less classical algebraic methods continues to be good and useful fun. British retired army officers used to be particularly clever at this sort of thing, and in some cases developed it into a very serious and scholarly hobby.

In modern and advanced probability theory there are some very powerful, general, and elegant methods which to a large extent replace the older algebraic methods. These newer techniques involve what is called *set theory*, *measure theory*, *generating functions*, etc. A considerable amount of serious study is necessary in order to understand and use these more powerful techniques. They are exceedingly interesting, and they are of great practical importance in the modern world, even in the modern world of business. For example, the November 1960 issue of *Journal of Research and Development*, of the International Business Machines Corporation, is entirely devoted to combinatorial problems. It includes an article the title of which, "The Enumeration of Trees by Height and Diameter," sounds like forestry! But it is in fact concerned with the enumeration of the characteristics of figures such as those in Fig. 11. These are all "rooted trees" with five "points" each, but of various "heights." The resemblance of the

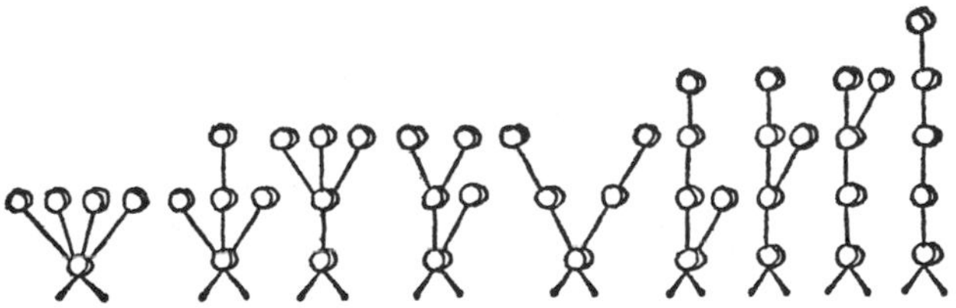

FIG. 11

diagrams to ordinary trees with branches is a little farfetched; but you can nevertheless see why mathematicians use these picturesque words for branching diagrams of this sort. These examples came from a modern book on enumeration,[8] written by a mathematician employed by the Bell Telephone Laboratories.

It is easy to see why "trees" of this sort may be useful in setting out, and enumerating, the various things that can happen under certain conditions. We already have had an example of this in Fig. 7, page 49. As another example, the following diagram, Fig. 12, (borrowed from the Kemeny book referred to previously) shows *half* of the tree of total possibilities when the Dodgers play the Yankees in a World Series. You of course remember that the Series is won by the team which first wins four games. In the figure a letter D stands for a Dodger win and a letter Y for a Yankee win, and the circles designate the ways the Series can end.

[8]John Riordan, *An Introduction to Combinatorial Analysis* (New York: John Wiley & Sons, 1958).

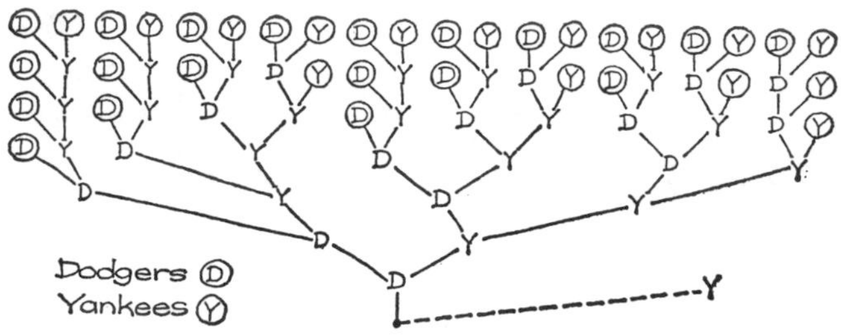

FIG. 12 *Half of the tree of logical possibilities for the outcomes of a World Series played between the Dodgers and the Yankees. The dotted line leads to the other half of the tree.*

This diagram is that half of the total "tree" which starts out with a Dodger victory. There are 35 different ways in which the series can terminate if the Dodgers win the first game, and (symmetrically) 35 more if the Yankees win the first. As the half-diagram implies, there is only one way in which the Dodgers could win in four games, whereas there are 20 ways in which the Dodgers can win in 7 games (the number of circles across the top row of the diagram).

We will conclude this chapter, which gives you only a glimpse of a large and fascinating field, with a few questions which, using the ideas and formulas above, you ought to be able to answer.[9]

1. Assuming that every inhabitant has just three initials, how many residents must a city have to make it inevitable that at least two residents have the same three initials?

2. How many straight lines can be drawn through seven points in such a way that each line passes through two of the points?

3. How many connections must a telephone

[9]The answers, in order, are given in a footnote on page 140.

exchange be able to set up if the exchange serves 100,000 subscribers?

4. How many three-letter words, each to contain one of the five regular vowels, can be made from an alphabet of 26 letters, permitting repetitions?

5. There are three different routes from Sparkesville to Hooperstown. How many ways can a round trip be made from Sparkesville to Hooperstown and back? How many ways that take a different return route?

6. How many types of banana splits can Howard Johnson's make with any two of twenty-eight sorts of ice cream and any one of ten syrup flavors?

7. How many poker hands of 5 cards can be dealt from a deck of 52 cards?

8. A man has two dozen shirts and wishes to take three with him on a weekend trip. How many combinations has he to choose from?

9. A sporting writer criticizes a coach for "not trying out every combination and then sticking to the best." Assume a football squad of 40 players, and suppose that the coaches work 24 hours a day every day of the year. Let each combination be tested for 5 minutes. How many years are required to carry out the suggestion?

Chapter V

SOME BASIC PROBABILITY RULES

It is a truth very certain that when it is not in our power to determine what is true we ought to follow what is most probable.

DESCARTES, *Discourse on Method*

A Preliminary Warning

In this chapter I want to set before the reader some of the most basic rules of probability theory. In old-fashioned books these rules would be discussed and "proved" in a way that almost any reasonably bright person would find less than satisfactory. In very modern books the discussion would be precise and compelling; there would be careful recognition of what is *assumed* and what is *demonstrated.* But these modern books would use notation and concepts (sample space, sample points, universal sets, sub sets, intersection, union, and disjunction of sets, the complement of a set, etc.) which are as neat as can be, which are highly useful and in fact essential for a serious modern treatment, but which nevertheless *do* seem strange and even formidable to persons who have not worked with them for some time, especially to persons who do not have a natural appetite for mathematics.

The real joker in that last sentence is the phrase "worked with them for some time." One cannot gain a friendly and relaxed acquaintance with an assortment of unfamiliar notation and new concepts just by looking at them a few times. He has to put pencil on paper, do examples, and gain familiarity with the new notation through actual use.

I do not assume that I have as tight control as that over the readers of this book. If you are *that* interested and *that* serious, you already have your money's worth, and you should drop this volume and start systematically on one of the several excellent textbooks now available.

So again I am going to compromise. By foreswearing the more technical modern procedures and terminology, I will sometimes lose in precision and in compactness of explanation. But at least I will try not to frighten you away.

Independent Events and Mutually Exclusive Events

If, of two events A and B, the one has no conceivable relationship with or influence on the other, then the two events can be called *independent* events. If the event A is "throwing a six with a die" and the event B is "cutting a deck of cards and obtaining a face card," then A and B are independent. The two procedures are clearly unrelated.

If the event A is "getting a head in the first toss of a certain coin" and the event B is "getting two successive heads by tossing the same coin a second time," then A and B are not independent events. For B cannot happen as a result of the two tosses unless A happens on the first toss. The two events are interlocked, and hence are *not* independent.

When events *are* related, there is one special kind of interrelation that deserves a separate name. If the dependency is of such a nature that if *A* occurs *B* simply cannot occur, whereas if *B* occurs *A* simply cannot, then the two events are called *mutually exclusive* events.

In probability discussions a person has to be very precise about the use of language. To illustrate this point, suppose that ten dice have been thrown, and consider the following description of various events which may have occurred. There may be:

1) A six, plus nine unspecified numbers.
2) Two sixes, plus eight unspecified numbers.
3) A six, plus nine numbers no one of which is a six.
4) At least one six.
5) No sixes at all.

The accurate description of Case 1 is to say that "there is at least one six." There may be more than one, for there may be one or more sixes among the nine unspecified numbers.

The accurate description of Case 2 is "at least two sixes." Case 3 is "exactly one six"; whereas Case 5 is clearly "no six."

Case 5 is completely distinct from the preceding four cases. Case 4 and Case 5, between them, cover all possibilities. Cases 1, 2, and 3 are all parts of, so to speak, Case 4.

It is useful and accurate to say that Cases 4 and 5 are mutually exclusive. When you throw any number of dice one of these two descriptions must apply, and only one of them does apply. If you "throw at least one six" then it is excluded that you "throw no six at all." And if you "throw no six at all," then it is excluded that you "throw at least one six."

This last remark is an important one. When you toss a die 10 times, all the things that could possibly happen divide up into two categories, in one of which "a six is tossed at least once" and in the other of which "no six is tossed at all." This leads at once to a more general statement, and one which is of frequent use: Namely, the complete complement of "doing something at least once" is "to do it not at all." Since one of these two mutually exclusive events is the complete complement of the other, the two combine to include all possible cases.

Thus if you roll a pair of dice a number of times, one of two mutually exclusive things can happen: You may get at least one double six, or you may get no double six. And what actually happens is bound to be one of the two.

This probably sounds so obvious that you wonder why I am stressing it. But rather frequently beginners in probability make the mistake of supposing that the opposite of *getting no double six* is *getting one double six. That is wrong.* The opposite of *getting no double six* is getting either one double six, or two double sixes, or three double sixes, etc.; that is to say, the opposite is *getting at least one double six* (or *getting one or more double sixes,* which is the same thing).

Converse Events

If we have N equally probable cases, only n of which result in an event E, then the remaining $N - n$ clearly form all the cases which do *not* lead to E.

Thus if $P(E)$ with the value

$$P(E) = \frac{n}{N}$$

is the probability for the event E, then

$$\frac{N - n}{N} = 1 - \frac{n}{N} = 1 - p$$

is the probability that E does not occur, or the probability, as one sometimes says, of *not* E. In modern notation *not* E is written $\bar{E}$.

There are under consideration here only two possible outcomes. Either the event E occurs, or it does not occur. One of these is the complete converse of the other. Add them together and you get all possible outcomes. In more formal, but exactly equivalent terms:

$$\left.\begin{array}{l}\text{Probability of the event } E = P(E) = p \\ \text{Probability of not } E = P(\bar{E}) = 1 - p\end{array}\right\} \qquad (7)$$

(Probability the event E occurs) + (Probability the event E does not occur)

$$\begin{aligned} &= (\text{Probability of all possible outcomes}) \\ &= p + (1 - p) \\ &= 1 \text{ or certainty} \end{aligned}$$

Fundamental Formulas for Total and for Compound Probability

Assume N equally probable cases[1] of a compound

[1]As you will recognize, we are about to develop some theory of a mathematical model in which the cases are strictly, by assumption, equally probable. We should remember this when we *apply* the theory.

event consisting of the simultaneous occurrence or nonoccurrence of two component events, which we will call A and B. (To illustrate, A might be "getting a head in tossing a coin" and B might be "getting a five when tossing a die." Then A can happen in two equally probable ways, B in six equally probable ways, the compound event of tossing them both simultaneously can occur in two times six or twelve equally probable ways, so that N would be twelve.)

As a matter of convenient notation, suppose

"A and B" means that A happens and that B happens,

"A and $\bar{B}$" means that A happens but that B does not happen, etc.

Furthermore (this sounds a little complicated, but it really is not) suppose that

A and B occurs in α of the total N cases,

A and $\bar{B}$ " " β " " " " " ,

$\bar{A}$ and B " " γ " " " " " ,

$\bar{A}$ and $\bar{B}$ " " δ " " " " " .

Since we have accounted for all possibilities,

$$\alpha + \beta + \gamma + \delta = N$$

Now, it is very easy, in terms of these quantities and in view of the fundamental definition (1), page 74, of probability, to write down eight different, interesting, and important probabilities.

As before, $P(A)$ means "the probability that A happens," $P(\bar{A})$ means "the probability that not-A happens — i.e., that A does not happen." And now we have to have a little new, but perfectly obvious notation. We will let $P(A \text{ if } B)$ have the natural meaning "the probability of A if B has in fact happened." Then, of course, $P(A \text{ and/or } B)$ means "the probability that A happens, or that B happens, or that they both happen," or, what is exactly the same thing, "the probability that at least one of A and B happens." Similarly $P(A \text{ and } B)$ obviously means "the probability that both A and B happen."

Then what is $P(A)$? From the above little table one sees at once that A happens in α cases (when both A and B happen) and in β cases (when A happens but B does not) of the N equally probable cases. Thus

$$P(A) = \frac{\alpha + \beta}{N}$$

In just the same way,

$$P(B) = \frac{\alpha + \gamma}{N}$$

$$P(A \text{ and } B) = \frac{\alpha}{N}$$

$$P(A \text{ and/or } B) = \frac{\alpha + \beta + \gamma}{N}$$

This last equation results from the fact that at least one of A and B happens in the α cases in which they both happen, in β cases in which A happens but B does not, and in γ cases in which B happens but A does not.

What is $P(A \text{ if } B)$? Well, the "if" condition means that we must restrict ourselves to the $\alpha + \gamma$ cases in which B happens, and we observe that A happens in α of these. Thus

$$P(A \text{ if } B) = \frac{\alpha}{\alpha + \gamma}$$

Similarly (and don't be lazy now: check each one for yourself!)

$$P(A \text{ if } \bar{B}) = \frac{\beta}{\beta + \delta}$$

$$P(B \text{ if } A) = \frac{\alpha}{\alpha + \beta}$$

$$P(B \text{ if } \bar{A}) = \frac{\gamma}{\gamma + \delta}$$

Although these equations are so essentially simple to derive, they are really very important and interesting.

For instance, using these values and a little elementary algebra we have

$$\begin{aligned} P(A) + P(B) &= \frac{\alpha + \beta}{N} + \frac{\alpha + \gamma}{N} = \frac{2\alpha + \beta + \gamma}{N} \\ &= \frac{\alpha}{N} + \frac{\alpha + \beta + \gamma}{N} \\ &= P(A \text{ and } B) + P(A \text{ and/or } B) \end{aligned}$$

Let's change the order a little and write

$$P(A \text{ and/or } B) = P(A) + P(B) - P(A \text{ and } B) \quad (8)$$

This equation says, and this is really a very neat and exciting result, that, faced with two possible events, the probability that one or the other or both (at least one) of them will happen is equal to the sum of the probabilities that each will happen, diminished by the probability that they both will happen. Would this have been really clear and convincing to you without the preceding algebra? If you think a bit you can see, on general grounds, why this statement is true. When you consider the cases in which

A happens, you are including the cases in which both A and B happen; and when you then add the cases in which B happens, you include a second time the cases in which they both happen. Clearly, you have to subtract the extra cases of both happening.

If the two events are mutually exclusive, the $P(A \text{ and } B) = 0$ since in these circumstances they *can't* both happen. Thus for two *mutually exclusive* events we have the formula

$$P(A \text{ or } B) = P(A) + P(B) \tag{9}$$

where we have written $P(A \text{ or } B)$ rather than $P(A \text{ and/or } B)$ because, A and B now being mutually exclusive, they both *can't* happen. This is a very fundamental formula of probability theory, and it states what is sometimes called the "law of total probability," namely, that the total (either-or) probability that one or the other of two mutually exclusive events will occur is the sum of their respective probabilities. *When two events are mutually exclusive*, their probabilities *add* to give the probability that one or the other will happen.

"Heads" and "tails" are mutually exclusive possibilities when you toss a coin once. What is the probability (using the result we have just derived) that when you toss a coin you get a head or a tail? It is $1/2 + 1/2 = 1$, or certain, as it jolly well ought to be!

If you toss a die, what is the probability that you get a three or a four? These are mutually exclusive, and the probability is $1/6 + 1/6 = 1/3$. And so on. . . . We will be making frequent use of this additive property of the probabilities of mutually exclusive events.

Going back to the eight equations we wrote down, we see that (inserting in both the numerator and denominator the factor $\alpha + \beta$, which cancels out and does not affect the correctness of the equality)

$$P(A \text{ and } B) = \frac{\alpha}{N} = \frac{\alpha}{\alpha + \beta} \cdot \frac{\alpha + \beta}{N}$$
$$= P(B \text{ if } A) \cdot P(A)$$

and that it also is true that,

$$P(A \text{ and } B) = \frac{\alpha}{N} = \frac{\alpha}{\alpha + \gamma} \cdot \frac{\alpha + \gamma}{N}$$
$$= P(A \text{ if } B) \cdot P(B)$$

Thus,

$$P(A \text{ and } B) = P(A \text{ if } B) \cdot P(B) = P(B \text{ if } A) \cdot P(A)$$

Now if A and B are independent events, then in the phrase "A if B," the part "if B" simply doesn't count or restrict at all, for the happening or non-happening of A does not depend at all on whether B has happened or not. That is to say, $P(A \text{ if } B)$ is exactly the same thing as plain $P(A)$. And similarly $P(B \text{ if } A)$ is, in these circumstances, of no interrelation whatsoever between A and B, the same as $P(B)$. Thus, *for independent events*

$$P(A \text{ and } B) = P(A) \cdot P(B) \tag{10}$$

or in words: *If two events are independent, the probability that they both will occur is the product of their respective probabilities.*

If the probability that I will fall down and break my leg, in some stated period, is 1 divided by 1,000,000, or 10^{-6}; and if the probability that the Yankees will win the pennant in the American

League next year is 1/3; and if the probability that it is going to rain tomorrow is 1/20, then the probability of occurrence of all three events, no one of which affects the others,[2] is the product of their respective probabilities.

When I toss a coin, what happens on any one toss is, ruling out any skill that would give me some degree of control, absolutely independent of what happened on the previous toss, and on all previous history of tossing. Thus what is the probability of throwing a head three times in a row? It is $1/2 \times 1/2 \times 1/2$, or 1/8. What is the probability of tossing three tails in a row? The same, 1/8. What is the probability of tossing a head, then a tail, then a head? Again the same value, 1/8. What is the probability of tossing any one specified order of heads and tails in three tosses? Again, 1/8.

Does this make sense? Well, how many distinct orders are there in three tosses of a coin? Remember the kind of reasoning we used in the preceding chapter. You can toss the first time in two ways. No matter how that comes out, you can toss the second time in two ways; and the third time in two ways. You can toss the three successive times in $2 \times 2 \times 2$, or 8 ways. They are all equally likely, so the probability of any one specified order ought to be 1 divided by 8. So, with another line of reasoning, we can check the result stated in the previous paragraph.

If we toss a coin n times, the probability of n successive heads, or of n successive tails, or of any

[2]Unless the rain makes it slippery, or a Yankee victory leads to too exuberant a celebration.

one specified pattern of heads and tails is—in all three cases—1 divided by 2^n.

If you remember the old story about paying for shoeing a horse, two cents for the first nail, four cents for the second, eight cents for the third and so on (2^n cents for the n'th nail), you recall how amazingly a number grows if you keep on doubling it. The probability of throwing ten heads in a row is $1/2^{10}$ or 2^{-10} or one chance in 1024. The chance of 20 heads in a row is less than one in a million.

We could go on almost indefinitely deriving formulas; but let's stop and have some fun with the ones we now have available.

Chapter VI

SOME PROBLEMS

The record of a month's roulette playing at Monte Carlo can afford us material for discussing the foundations of knowledge.

KARL PEARSON

Foreword

This chapter starts out with a rather detailed discussion of one of the de Méré problems which initiated the real mathematical theory of probability. Then there is an account, in detail for the simpler ones and sketchily for the others, of several of the classical problems which characterized the first century of the development of the theory. Finally, at the end, there is a small collection of problems that you, the reader, may wish to try your own hand at.

You could perfectly well read the rest of the book without wading all the way through this chapter. If you dislike equations — even a very simple one gives my wife vertigo — then skip to the next chapter.

But I hope that you will find yourself coming back to this chapter, to pick up any pieces you dropped the first time. For these are really fascinating problems.

The First Problem of de Méré

You will recall that the first problem de Méré considered involved two games, in the first of which the gambler throws a die four times, the house betting even money that the gambler will get at least one six. What, in fact, is the probability of getting at least one six in four throws of a die?

The way to go about calculating this, as we indicated in a preliminary way on page 106, is to go about it indirectly — to attack from the rear, so to speak. We therefore start not by calculating the probability of the event E, defined as obtaining at least one six in four throws of a die, but by doing the easier job of calculating the probability of the complementary event $\bar{E}$, defined as not getting *any* sixes in four throws of a die.

Not getting a six in four throws of a die is a compound event composed of four independent component events — "not getting a six on the first throw," "not getting a six on the second throw," "not getting a six on the third throw," and "not getting a six on the fourth throw." The probability of each one of these four component events is clearly 5/6, since there are, among the six equally probable outcomes, five outcomes — namely the throwing of 1, 2, 3, 4, or 5 — each of which results in the event (no six) in question. Since what happens on the second throw is completely independent of what happened on the first, etc., we have by the basic rule (10) on page 111,

$$P(\bar{E}) = p = \frac{5}{6} \times \frac{5}{6} \times \frac{5}{6} \times \frac{5}{6} = \left(\frac{5}{6}\right)^4$$

But then, as we saw back on page 111,

$$P(E) = 1 - p = 1 - \left(\frac{5}{6}\right)^4$$

is the probability we really wanted, of getting at least one six in four throws of a die. If we multiply out, we have

$$1 - \frac{625}{1296} = \frac{1296 - 625}{1296} = \frac{671}{1296}$$

Now since half of 1296 is 648, we have found ourselves with a fraction slightly greater than 1/2. In other words, when throwing a "true"[1] die four times, there is a little higher chance of throwing at least one six than of not doing so. Of the 1296 equally probable outcomes when throwing a die four times, 671 involve throwing at least one six, whereas the remaining 1296 − 671 = 625 outcomes, etc., do not involve any six at all. Stated in terms of odds, the odds are 671 to 625 that the gambler loses, and the house wins.

With this first example explained in detail, we can more rapidly analyze the second game of this problem of de Méré.

What is the chance of at least one double six in twenty-four throws of two dice? Again we go at it backwards.

The probability of not getting a double six in one throw is 35/36. The probability of not getting a double six in twenty-four throws is 35/36 multiplied by itself until there are 24 factors — or as one

[1]The word "true" here implies the transition from the pure theory of the mathematical model to the real case of a die made and rolled so fairly that we can believe that the theory applies.

says in algebra, "raised to the 24th power." Subtracting this from 1 (certainty) gives the desired probability of the complementary event — throwing at least one double six. This is,

$$1 - \left(\frac{35}{36}\right)^{24}$$

which equals,[2] to four decimals, 0.4913.

This is a little *under* 1/2, so the gambler has a little less than an even chance of losing — a little better than an even chance of winning. The game (as de Méré seems to have satisfied himself in some way) is slightly against the house.

If we make the same calculation for twenty-five throws instead of twenty-four, the probability of at least one double six — the probability that the house wins — then turns out to be 0.5054, which is very slightly more than 1/2 — very slightly favorable to the house.

The old gambling rule, of which we spoke in Chapter II, gives the result 24 as the smallest number of throws favorable to the house. The rule was in error, since, as we have just seen, the correct answer is 25.

The old gambling rule in question was first given a sound theoretical basis by the mathematician Abraham de Moivre, in a famous book *The Doctrine of Chances*, published in 1716. Although we will have to be sketchy about one detail, we can easily establish de Moivre's form of the rule.

For if p is the probability that a certain event will happen on one trial, then $(1 - p)^n$ is the probability,

[2]It would be an awful bore to multiply this all by hand. One uses logarithms or some sort of a computing device.

as we have seen, that it will not happen at all in n trials, and hence

$$1 - (1 - p)^n$$

is the probability that it will happen at least once in n trials. The break-even point, in gambling, occurs when this probability is 1/2. So the "critical number of trials" is the integer next larger than the number n (which in general will not come out to be a whole number) which satisfies the equation

$$1 - (1 - p)^n = \frac{1}{2}$$

Rearranging terms,

$$(1 - p)^n = \frac{1}{2}$$

To solve this kind of equation, in which the unknown is an exponent, one takes the logarithm of both sides. (You learned how to do this in algebra, remember?) The result is, changing signs on both sides,

$$-n \log (1 - p) = \log 2$$

Now comes the sketchy point, for you now have to take something on faith. In calculus it can be rather easily shown that[3]

$$-\log (1 - p) = p + \frac{p^2}{2} + \frac{p^3}{3} + \frac{p^4}{4} + \cdots$$

[3]For simplicity in the expressions, the logarithms used here are the so-called "natural" logarithms, using the base e (the same old e you saw in expression (A) on page 90) rather than the ordinary logarithms, which use the base 10. You needn't worry about this.

this equation being valid for values of p between 0 and 1 (just the kind of values p has in probability problems!), and the right-hand side being more and more accurate as one takes more and more terms.

If p is a small number, then p^2 is much smaller, p^3 awfully small, and so on. In other words, for pretty small p we can disregard all the terms on the right except the first one, to obtain the approximate formula

$$np = \log 2$$

$$n = \frac{\log 2}{p} = \frac{.6931}{p}$$

This formula will in general not produce an integer, and thus the *critical number* will be the first integer larger than $.6931/p$. If we retained the other terms in the above series for the negative logarithm of $(1 - p)$, then the formula would read,

$$n = \frac{.6931}{p + \frac{p^2}{2} + \frac{p^3}{3} + \cdots}$$

Since in our approximate formula we dropped all the terms except p, we have reduced the denominator in value, and have thus increased n above its correct value. That is, the de Moivre gambling rule always gives *somewhat too large* a value for n.

That may, or may not, cause a real error. If the true value of n were in some case 8.31 and the approximate formula gave 8.57, that would not hurt a particle, since either $n = 8.31$ or $n = 8.57$ would lead to the same choice of the critical number, namely 9, as the next higher integer. But if the correct value were 4.96 and the formula gave 5.03, that would be bad. For the correct value 4.96 leads to 5 as the (correct) critical number, whereas 5.03 would force one up to 6 as the critical number, and would thus give an inaccurate result.

In the case of a double six with one throw of two dice, p has the rather small value 1/36, and the approximate formula gives $n = (36)(.6931)$ or 24.95, which indicates that the odds shift from *unfavorable* on 24 throws to *favorable* on the least integral number larger than 24.95, namely on 25 throws. And we have found already that 25 is in fact the number of throws at which point the odds shift to a situation favorable to the house.

In other words, the approximate rule expressed by

$$\text{critical number is the first integer larger than } \frac{.69311}{\text{probability on a single trial}} \qquad (11)$$

works perfectly well for the de Méré problem!

But we said, earlier, that de Méré used the old gamblers' rule, and it did *not* work, giving the incorrect number 24 instead of the correct 25.

It was not, however, the de Moivre rule we have just derived which de Méré used, but an older and less accurate form. This older form, the real old gamblers' rule, went as follows: If in a first game the probability of success on a single trial is p_1 and the critical number is n_1, then for a second game, in which the probability of success is p_2, the critical number n_2 will be

$$n_2 = \frac{p_1}{p_2} \times n_1 \text{ (Old Gamblers' Rule)} \qquad (12)$$

In the de Méré problem, in the first game n_1 was known to be 4 and p_1 (the probability of a six on one throw of a single die) was 1/6. For the second game p_2 (the probability of a double six on one throw of a pair of dice) is 1/36. Thus the old rule gave

$$n_2 = \frac{36}{6} \times 4 = 24$$

which was wrong.

What is the relation between de Moivre's gamblers' rule (as given by Equation 11), and the Old Gamblers' Rule (12)? Write down the de Moivre rule for game one,

$$n_1 = \frac{.6931}{p_1}$$

and for game two,

$$n_2 = \frac{.6931}{p_2}$$

and then divide one of these equations by the other. The result is

$$\frac{n_1}{n_2} = \frac{p_2}{p_1}$$

or

$$n_2 = \frac{p_1}{p_2} \times n_1$$

which looks just like the Old Gamblers' Rule (12). But in this last equation, the n_1 on the right-hand side is not, as in the Old Gamblers' Rule, the critical number for the first game (which number is of course an integer), but is the value

$$n_1 = \frac{.6931}{p_1}$$

given by the de Moivre formula. If we use the

de Moivre formula for the first de Méré game, concerned with getting at least one six in a certain number of throws with one die, we get

$$n_1 = \frac{.6931}{1/6} = 4.1586$$

This is, as is always the case with this formula, *too big* a value; and in this case it is much too big and would falsely indicate 5 as the critical value.

If we kept one more term in the series we would have

$$n_1 = \frac{.6931}{1/6 + 2/36} = \frac{.6931}{8/36} = 3.1189$$

which is considerably smaller and leads to the correct critical value 4.

You can now see the joker. The de Moivre formula does not work reliably unless p is small. In the first de Méré game $p = 1/6$, which is a larger probability than the formula can tolerate, and the formula leads to the incorrect critical number 5 (the first integer larger than 4.1586) instead of the correct result 4. In the second game, $p = 1/36$ is small enough so that the de Moivre formula works well.

The Old Gamblers' Rule is thus no good as a basis for comparing two games unless p is small enough *in both games* to make the de Moivre rule work satisfactorily. When p is small enough in both cases, the de Moivre rule, applied to the two games and then divided, produces the Old Gamblers' Rule.

The conclusion is: Don't use the Old Gamblers' Rule except in comparing two games in both of which p is small — at least as small as 1/10 and preferably smaller. And actually it is much simpler in any case to use the de Moivre rule (11) or the improved form, which works for somewhat smaller p, namely,

$$\text{critical number is the first integer larger than } \frac{.6931}{p + p^2/2} \text{ or } \frac{1.3862}{2p + p^2}$$

where p is the probability on a single trial.

Or still better: Don't use any gamblers' rule and don't gamble! In a later chapter we will see that this is sound scientific advice, apart from any moral consideration.

We have spent a lot of time and fuss on this unimportant problem. But it is the first problem we have worked out together, so it is a good idea to do a thorough job of it. And, after all, this is the problem which started off the mathematical theory of probability.

The Problem of the Three Chests

Suppose you are standing in front of three chests which look exactly alike, each of the chests containing two drawers. We will call the chests A, B, and C. Each drawer of Chest A contains a gold piece. In Chest B one drawer has a gold piece, one a silver piece. In Chest C each drawer contains a silver piece.

Choose a chest at random, and then choose at random one of its two drawers. To choose "at random" means that you are just as likely to choose one as another, which is reasonable since the chests look exactly alike. The various choices are thus equally likely, and the probability of choosing Chest A is 1/3. Similarly, the probability of choosing one drawer is 1/2, and of choosing the other is 1/2 also.[4]

[4]Again we are talking about real chests, but calculating an idealized model. We won't go on saying this each time.

What is the probability of choosing a particular chest and a particular drawer in that chest? This is a compound event with two independent subevents, and so the probabilities multiply and the answer is $1/3 \cdot 1/2 = 1/6$.

If you choose a chest at random and one of its two drawers at random, what is the probability of getting a gold piece? There are three mutually exclusive things that can occur in the choice of chest: You pick A, or you pick B, or you pick C. If you pick A (the probability of which is 1/3) then you are certain (probability is 1) of getting gold. Thus the compound probability of picking A and of getting gold is $1/3 \cdot 1$. Similarly, thc probability of picking B and of getting gold is $1/3 \cdot 1/2$. The probability of picking C and of getting gold is $1/3 \cdot 0$. Thus the total probability of one or another of these three mutually exclusive events (remember from Chapter IV that the either-or probabilities of mutually exclusive events *add*) is

$$\frac{1}{3} \times 1 + \frac{1}{3} \times \frac{1}{2} + \frac{1}{3} \times 0 = \frac{1}{3} + \frac{1}{6} = \frac{1}{2}$$

That makes very good sense. You ought to have an even chance of getting a gold piece, because, after all, half are gold and half silver and they are equally accessible to your choice.

Now (and this is much more interesting) choose a chest at random, choose one of its drawers at random, and open the drawer. Suppose it contains a gold piece. What is the probability that the other drawer in this same chest also contains a gold piece?

Having observed that one drawer does contain a gold piece, we definitely eliminate Chest C. The chest chosen must be either A or B. It might be asserted that, since the chest selected is either A or B, and since the original choice of chest was a completely free and unbiased one, the probability is 1/2 that the chest selected is A, and also 1/2 that it is B.

But this is incorrect! For the gold piece which has been revealed is one of *three* gold pieces, which were all *equally accessible to original choice.* Namely, it may be gold piece number one in Chest A, gold piece number two of Chest A, or the single gold piece in Chest B. *These* three are equally probable. In two of these three cases the other coin is gold. And therefore the answer to the question, "What is the probability that the other drawer in the selected chest also contains a gold piece?" is that this probability is 2/3.

This is an old and classical problem; and it shows very clearly that you have to be careful, in figuring a probability, to deal with events that are *in fact* equally probable.

You can make a profitable game out of this problem. Instead of chests and gold and silver pieces, take three identical cards. Make a red mark on both sides of one, a black mark on both sides of the second, and mark the third black on one side and red on the other. Mix them up in a hat, pick out a card at random, and put it down on the table without disclosing to yourself or anyone else what color is marked on the concealed side.

Suppose the upper side of the card is marked red. You say to your opponent, "Obviously we are not dealing with the black-black card. That one is clearly eliminated. We definitely have either the red-black card or the red-red card. We shuffled fairly and drew at random, so it is just as likely to be one of these as the other. I will therefore bet you even money that the other side is red."

It isn't too hard to find takers, although (as you

now see) the odds in favor of your bet are *not* even, but are actually two to one! The catch, of course, is the clause, "so it is just as likely to be one as the other." It is *twice* as likely that it is the red-red card! Forty years ago, when graduate students had to work for their living, the author used to teach this particular problem, at reasonable rates and using the experimental method, to his college friends.

A Few Classical Problems

The two problems we have considered so far in this chapter are classical ones, going back to the early history of probability theory. It is instructive, and I think interesting, to look at a few more classical problems to get the flavor of those beginning days.

Even before Lady Luck was born in 1654 there were some preliminary stirrings. Rabbi Ben Ezra, in the middle of the 12th century, made some calculations about permutations and combinations for astrological purposes, and it is easy to believe that he had some sort of probability notion in the back of his head.[5] And as we have mentioned earlier, a remarkable Italian named Girolamo Cardano wrote a manuscript entitled *Liber de Ludo Aleae* (*Book of Dice Games*). It was not printed until 1663. It contained chapters "On the Cast of One Die," "On the Cast of Two Dice," "On the Cast of Three Dice," "On Card Games," "On Luck in Play." One chapter raised the question, not yet settled, "Do those who teach play well?" Toward the end of his treatise there was a chapter "On Far-reaching Plans, Judgment, and Procedure." This last-named

[5]My friend, Professor Fred Mosteller, of Harvard University called my attention to this.

chapter referred, to be sure, to a game (backgammon). But the new ground had been broken; the book as a whole dealt not with what *must* happen, but with what is likely to happen.

It was some gamblers who later started Galileo Galilei thinking along these lines when they asked him why it was that a throw of three dice turns up a sum of 10 more often than a sum of 9. That is a question you can now perfectly well think out for yourself; and you will discover that of the 216 equally probable outcomes when three dice are thrown, 27 lead to a sum of 10, and only 25 lead to a sum of 9.

In 1770 there was published *Daniel Bernoulli's Problem*, Bernoulli being one of a famous family of mathematicians. Suppose you have a jar which contains n white balls, a second jar which contains n black balls, and a third which contains n red balls. (See Fig. 13.) Take at random one ball out of each jar. ("At random" doesn't mean much on this first round, but it will later.) Whatever you took (white, of course) out of Jar 1, put it in Jar 2; whatever you took out of Jar 2, put in Jar 3; whatever you took out of Jar 3, put in Jar 1. Now repeat this operation x times. What are the expected[6] numbers of the various colors of balls in each of the three jars?

I think you can sense that this problem, although phrased in terms of an artificial game with jars and colored balls, is in fact related to general and important questions as to *how things get mixed* when chance processes are at work.

[6]The exact meaning of this word in probability theory will be explained in the next chapter. But it is permissible here to think of the word in its usual sense.

FIG. 13

This is not an easy problem. Daniel Bernoulli solved it, and subsequent mathematicians, including Laplace, solved generalizations of it. The expected number of white balls, for example, in Jar 1 after the x drawings and replacements is

$$\frac{n}{3}\left[\left(1-\frac{1}{n}+\frac{\alpha}{n}\right)^x+\left(1-\frac{1}{n}+\frac{\beta}{n}\right)^x+\left(1-\frac{1}{n}+\frac{\gamma}{n}\right)^x\right]$$

where α, β, and γ are the three cube roots of unity.[7]

[7]You may be surprised at the phrase, "the three cube roots of unity." Perhaps you always thought that 1 was the only number that, when multiplied by itself two times so as to have three factors, would produce unity. But the cubic equation $x^3 = 1$, or $x^3 - 1 = 0$, has *three* answers just as any cubic equation does. Since $x = 1$ is obviously one of the answers, $(x - 1)$ must be a factor of $x^3 - 1 = 0$. If you divide out this factor $(x - 1)$ you have left the quadratic equation $x^2 - x + 1 = 0$. If now you solve this quadratic by the formula you learned in algebra, you find that the other two roots are the complex numbers

$$-\frac{1}{2}+\frac{\sqrt{3}}{2}i \quad \text{and} \quad -\frac{1}{2}-\frac{\sqrt{3}}{2}i, \text{ where, of course, } i = \sqrt{-1}.$$

If you think these cube roots of 1 are fishy, just multiply up and be convinced, remembering that $i^2 = -1$ and $i^3 = -i$.

Another classic problem, which we mentioned in Chapter II as the second of the two problems which de Méré posed to Pascal, is the *Problem of Points.* Many of the great mathematicians worked on more and more complete solutions to more and more general forms of this problem.

In a general form, the problem is this: Two players put up stakes and play some game, with an agreement as to what constitutes winning. Some interruption requires them to stop before either has won, and when each has some sort of a "partial score." How should the stakes be divided? A fair answer seems to be that each player's proportion should depend upon the probability that he would win if the game were continued. In the simple form of the problem, it is assumed that the two players have an equal chance of winning any single point.

There are many, many special forms to this problem. I give here only a couple of results obtained by Pascal.

If each player puts up a stake A, if $n + 1$ points are required to win, if the first player has n points and the second player none, if they stop then, the first player is entitled to

$$2\mathrm{A} - \frac{\mathrm{A}}{2^n}$$

If the circumstances are as before, but the first player has one point and the second player none, then the first player is entitled to

$$\mathrm{A} + \mathrm{A}\,\frac{1 \times 3 \times 5 \times \cdots (2n - 1)}{2 \times 4 \times 6 \times \cdots \quad 2n}$$

You will probably be content just to take an interested look at these results, for the *Problem of Points* is, in fact, a reasonably difficult one.

Another classic problem is the Problem of *Duration of Play*, first proposed by Huygens in 1657, solved by James Bernoulli and published eight years after his death, in 1713. The problem was treated in more general terms by several later mathematicians.

The first example of the Duration problem, as stated by Huygens, was this: A and B have 12 coins each. They play for these coins in a game with three dice. Whenever 11 is thrown (by either: it doesn't matter who rolls the bones), then A gives a coin to B. Whenever 14 is thrown, B gives a coin to A. The person who first wins all the coins wins the game.

Huygens stated that A's chance is to B's as 244,140,625 is to 282,429,536,481. (You can see that he had to work out some rather lengthy enumeration of the various outcomes.)

James Bernoulli took a more general case. He supposed A to have m coins and B to have n. They play a game in which their respective chances for any individual point are as a to b. The loser on each point gives a coin to the other. What is the chance, for A and for B, that he will win all the coins of his adversary?

These problems are called "duration of play" because the answer gives an indication of how long it is likely to take one to win and so finish the game. If the probability that one will wipe out the other is, for example, 0.9, then the game won't last long. If the probability that either one will wipe out the other is small, then the game obviously could go on and on.

This problem is obviously and directly relative to the question of being ruined when gambling. What is the probability that the gambler's adversary (the "house") will take all his coins? If so, then he is ruined, and cannot continue play.

Answer: For Bernoulli's case, defined above, A's chance of winning all of B's coins is

$$\frac{a^n(a^m - b^m)}{a^{n+m} - b^{n+m}}$$

In 1760 Daniel Bernoulli tried to calculate, by probability theory, the mortality caused at various ages by smallpox. Remember that health records did not exist at that time. His result was not of itself too interesting or important, for he was not in a position to invent a mathematical model that promised to be significantly related to the real facts. But it is very interesting to know that even at that early date probability theory was beginning to get out of the gambling halls.

Even earlier, Nicolas Bernoulli (another of the bright Bernoullis) was concerned with applying probabilistic ideas to the numbers of girls and boys born. Daniel speculated on the duration of marriages! This case is specially interesting, for he had the courage to propose a theoretical model for so complex a human phenomenon. In fact, he started from a pure game. (I suppose someone has written a song, "Love Is a Game.")

Suppose that a jar contains $2n$ cards, two of them marked 1, two marked 2, two marked 3, etc. Draw out m cards at random. What is the number of pairs one is most likely to find still in the bag? The answer is

$$\frac{(2n - m)(2n - m - 1)}{2(2n - 1)}$$

Suppose now that 500 men of a given age marry 500 women of the same age. Suppose the mortality table shows how many individuals die each succeeding year (i.e., are "drawn out of the bag"). How many marriages persist each year (i.e., "how many

FIG. 14

pairs are left in the bag")? Bernoulli assumed that his probability calculation applied directly. He then generalized his problem to cover the case that the husband and wife were not the same age.

The Birthday Problem

Suppose there are n people in a room. What is the probability that at least two of them share the same birthday — the same day of the same month? In our mathematical model we assume that a person is just as likely to be born on one day as another, and we ignore leap years, so that we always have 365 days in a year.

The event that at least two persons share the same birthday is complementary to the event in which they all have distinct birthdays. We start with one person. Whatever day it may happen to be, he *has* a birthday. The probability that person number two has a different birthday is clearly,

$$\frac{364}{365}$$

since it will be different if he was born on any one of the 364 days remaining after we cross off, so to speak, the birthday of the first person. When we advance to the third person, there are 363 permissible days left, so the probability that the third person's birthday differs from that of the first and the second is

$$\frac{363}{365}$$

These are independent events, so the compound probability that number two differs from number one and that number three differs from both number one and two is

$$\frac{364}{365} \times \frac{363}{365}$$

In just the same way the probability of all distinct birthdays for the first four persons is

$$\frac{364 \times 363 \times 362}{365 \times 365 \times 365} = \frac{365 \times 364 \times 363 \times 362}{365 \times 365 \times 365 \times 365}$$

where, in the last-written form, we have inserted an extra 365 both upstairs and downstairs in the fraction (leaving its value unchanged, of course) just so the number of factors will, for convenience in remembering, be equal to the number of persons to which the expression is applied.

It is now easy to generalize to the formula for the probability of all distinct birthdays for n persons. It obviously is

$$\frac{365 \times 364 \times 363 \times 362 \cdots}{365 \times 365 \times 365 \times 365 \cdots}$$

with n factors in both the numerator and denominator of the fraction.

Therefore the probability of the complementary event, namely that at least two persons share the same birthday, is

$$1 - \frac{365 \times 364 \times 363 \cdots (365 - n + 1)}{365^n}$$

where, in this final form, we have written down the explicit form of the n'th factor in the numerator. (Maybe you had better check up, by adopting some reasonably small value for n, that the n'th factor is really $365 - n + 1$ and not $365 - n$, which you might have wrongly guessed.)

Having worked out this neat general formula, let's look at the numerical results it gives for a few interesting values of n.

When there are 10 persons in a room together, this formula shows that the probability is 0.117 (or better than one chance in nine) that at least two of them have the same birthday. For $n = 22$ the formula gives $p = .476$; whereas for $n = 23$ it gives $p = 0.507$. So if there are 23 persons in a room, there is better than an even chance that at least two have the same birthday.

Most people find this surprising. But even more surprising is the fact that with 50 persons, the probability is 0.970. And with 100 persons, the odds are better than three million to one that at least two have the same birthday.

In World War II, I mentioned these facts at a dinner attended by a group of high-ranking officers of the Army and Navy. Most of them thought it incredible that there was an even chance with only 22 or 23 persons. Noticing that there were exactly 22 at the table, someone proposed we run a test. We got all the way around the table without a duplicate birthday. At which point a waitress remarked, "Excuse me. But I am the 23rd person in the room, and my birthday is May 17, just like the General's over there." I admit that this story is almost too good to be true (for, after all, the test should succeed only half of the time when the odds are even); but you can take my word for it.

Montmort's Problem

We will take a brief look at a rather more general problem that also deals with coincidences. This problem was first proposed by the mathematician Montmort in his treatise on the analysis of games of chance, printed in 1708.

Suppose you have a jar containing n balls numbered $1, 2, 3, \cdots n$, respectively. They are well mixed and then drawn one at a time. What is the probability that no ball is drawn in the order indicated by its label? Or to rephrase the question, suppose that as one draws out the balls, one at a time, he counts out loud, saying, "one $\cdots$ two $\cdots$ three $\cdots$," etc. What is the probability that there will be no coincidence — no drawing of a ball bearing the number just called out?

Since in the original form the jar was supposed to contain 13 balls, this game was called *Treize*, the French word for thirteen. It is also called *Rencontres*, or *Coincidences*, to use the English word; and is perhaps best called Montmort's Problem, since he was the originator of the theory for the game.

The total number of equally likely outcomes is easy to reckon. It is simply the total number of orders in which the n balls can be drawn out, one at a time; which is clearly the number of permutations of n distinct objects or $n!$ But the number of favorable cases — the number of orders of the balls in which no ball finds itself in its "natural" position (as ball No. 5 in position No. 5 $\cdots$) turns out to be rather difficult to figure out, as is characteristically true of problems in which you want to know the number of ways in which objects can be arranged, subject to special conditions.

The detailed solution of Montmort's problem is rather beyond our scope here, so we will just write the answer down. The probability of no coincidence with n balls is

$$p = \frac{1}{2!} - \frac{1}{3!} + \frac{1}{4!} - \frac{1}{5!} + \cdots$$

where one writes down, on the right side of the equation, $n - 1$ terms in all.[8]

Now this is a very remarkable and interesting

[8]The last one of these terms is $(-1)^{n-1}/n!$: but this expression looks a little complicated, and it's easier just to say, "Write down $n - 1$ terms."

result. We have already noticed that factorial n grows very rapidly, as n gets larger. So the successive fractions in this expression get *small* very rapidly. Moreover the sign moves back and forth, so that you *add* a little, then *subtract* less, then *add* still less, etc. This means that after just a few terms, the value of the right side settles down closer and closer to some value, and we can symbolically represent it as in Fig. 15.

FIG. 15 *The Convergence of an Oscillating Series*

In fact, look at the actual value of a few successive terms — and the accumulative sum (taking into account algebraic sign, of course):

TABLE V

THE PROBABILITY OF NO COINCIDENCES IN n TRIALS

Term	*Value of Term*	*Value of n*	*Accumulative Sum*
First	+0.500000 · · ·	2	0.500000 · · ·
Second	−0.166666 · · ·	3	0.333333 · · ·
Third	+0.041666 · · ·	4	0.375000 · · ·
Fourth	−0.008333 · · ·	5	0.366666 · · ·
Fifth	+0.001388 · · ·	6	0.368054 · · ·
Sixth	−0.000198 · · ·	7	0.367857 · · ·
Seventh	+0.000025 · · ·	8	0.367882 · · ·
Eighth	−0.000003 · · ·	9	0.367879 · · ·
Ninth	+0.000000 · · ·	10	0.367879 · · ·

The rather surprising fact is that the accumulative

sum — which is the probability of no coincidence — doesn't really change much as one considers problems with a larger and larger number of balls, once the number of balls has reached three or four! The chance of no coincidence with 4 balls differs from the chance with, say, 10 balls by only about 2 per cent. And the difference between 5 balls and 10 balls is less than 4/10ths of 1 per cent.

As one considers still larger numbers of balls, moreover, it is obvious that the probability of no coincidence changes practically not at all. To six decimals it is 0.367879 whether the number of balls is nine, or ten, or one hundred, or ten billion!

It is not difficult to see that this could be the case. With more and more balls the chance of coincidence on any one draw obviously gets smaller and smaller; but to compensate, the number of chances *for* coincidences keeps growing as n grows. Thus these two effects would tend to cancel each other out. That they indeed do, and so perfectly, is substantiated by the accurate and complete analysis.

The expression written above for the probability of no coincidence is in fact a series, which one learns how to derive in calculus, for $1/e$, where e is (again we meet this important number) the base of the natural, or Naperian, system of logarithms. The value of e is $2.71828 \cdots$ and the value of $1/e$ is $0.367879 \cdots$, checking the probability values written above. Since e is an important number, there has always been interest in calculating it to a large number of decimal places.[9]

It has recently been popular to pose the following

[9]See note at the end of this chapter.

problem: Suppose n men, attending a banquet, check their hats with a completely bird-brained checkroom girl who puts no numbers on the hats. The men all get drunk at the banquet, and as a result of all these happy circumstances, each man, when he leaves, gets (and, being drunk, accepts) a hat at random. What is the probability that, in spite of all this mix-up, at least one man gets his own hat?

This is exactly equivalent to Montmort's original problem of matching the numbered balls with the numbers called out as the balls are drawn. The men appear at the checkroom one at a time in sequence, and each gets a hat from a "shuffled" pile of hats. The probability that no man gets his own hat is $1/e$. The probability that at least one man does get his own hat is an event complementary to the event "none." So the probability is

$$1 - \frac{1}{e} = 1 - 0.367879 \cdots = 0.632121 \cdots$$

This result holds accurately so long as there are at least eight men at the party, and is approximately true for smaller number of guests.

Put otherwise, the odds are about 63 to 37 (or nearly two to one!) that at least one guest gets his own hat — and this rather astonishing result holds for a small number of guests, or for a huge party.

Suppose you take a deck of cards, shuffle it thoroughly, and then turn up the cards, one at a time, counting, "Ace, two, three, four, five · · · ten, Jack, Queen, King, Ace, two, three, four · · ·" What is the chance that you will "call" the identification of at least one card as you turn it up? Although this is a slightly different problem, because of the four suits, you will probably not be surprised to know that the odds here, also, of getting at least one "coincidence" are 63 to 37. If you pick out the thirteen cards of one suit, and play the game with them, then you are playing the classical *Treize*. You frequently can find a friend (a temporary friend) who will bet you even money that there will be no coincidence. He will lose his bet in about two out of every three games.

Try These Yourself

This is not a textbook; and you will not find a long list of problems that you must work out yourself. But at this stage you may enjoy trying your own hand at a few problems — or call them puzzles if that sounds more attractive.

They necessarily will be pretty artificial little problems, for we have to get behind us a couple more chapters of development of the subject before we will be in a position to comment on questions such as those back on pages 27 and 28 of Chapter I.

(a) Toss two coins. What is the probability that there will be at least one head? You know now that

Here are the answers, in order, to the nine problems on pages 100–1.

First problem: $26^3 + 1 = 17{,}577$ residents. Second problem, assuming that no three of the points are co-linear: $\binom{7}{2} = 21$. Third problem: $\binom{100{,}000}{2} = 4{,}999{,}950{,}000$. Fourth problem: $5 \cdot 26 \cdot 25 + 26 \cdot 5 \cdot 26 + 26 \cdot 26 \cdot 5 = 10{,}110$. Fifth problem: 9, 6. Sixth problem: 7560. Seventh problem: $\binom{52}{5} = 2{,}598{,}960$. Eighth problem: $\binom{24}{3} = 2024$. Ninth problem: about 22,000 years if the coaches test only the different combinations. If they also test each player in each position in every combination it would take over eight hundred million years.

this is easy, and the answer is 3/4. Does it comfort you to know that d'Alembert, a distinguished mathematician of the 18th century, thought that the answer was 2/3? He didn't correctly understand, as you do, the necessity of counting all the equally likely outcomes. He would have been a pushover for the bet on the three chests.

(b) A tosses 3 coins. B tosses 2 coins. What is the probability that A throws more heads than B? In this compound event the first component event can occur in 8 equally probable outcomes. The second has 4 equally probable outcomes: so the compound event has 32 equally probable outcomes. Of these . . .

(c) The Game of Snap. Shuffle two similar decks of n distinct cards each, and place each deck face down. Simultaneously turn over and remove the top card of each deck, observing what the two cards are. Repeat this process until the decks are exhausted, or until one exposes, at one turn, the same card on each deck. This latter event is called a "snap."

This turns out, upon reflection, to be exactly the same as Montmort's Problem; as is clear at once if you just suppose that, after being shuffled, the first deck is numbered 1 to n, down from the top card, and the second deck has the same numbers put on the same cards. So we know the answer. For a pack of one card the probability of snap is (of course) 1.0. For a two-card deck it is 0.5. For a five-card deck it is 0.633333 $\cdots$ and for a ten-card deck (or any larger deck) it is 0.63212 $\cdots$.

For a deck of 52 cards, the probability of snap is increased (by only a very small amount) if one either removes one card from each deck, or adds one card (say the Joker) to each deck. Why?

(d) Now let us, finally, consider a multiple problem,

the early part of which is very easy, but the final part of which will teach us something new.

A jar contains 20 white balls and 30 black ones. Check, by reasoning out for yourself, that if one ball is drawn, the probability[10]

of drawing a white ball is 2/5,

of drawing a black ball is 3/5.

If two balls are drawn in succession, then the probability

of drawing two white balls is $\frac{20}{50} \times \frac{19}{49}$

of drawing two black balls is $\frac{30}{50} \times \frac{29}{49}$

of drawing one white ball and then one black ball is

$$\frac{20}{50} \times \frac{30}{49}$$

of drawing one black ball and then one white ball is

$$\frac{30}{50} \times \frac{20}{49}$$

If two balls are drawn in succession, then replaced, and two are again drawn in succession, the probability of drawing first one white ball and then one black ball both times is

$$\frac{20}{50} \times \frac{30}{49} \times \frac{20}{50} \times \frac{30}{49} = \left(\frac{20.30}{50.49}\right)^2$$

The probability of drawing first three black balls and

[10]Assuming that any balls drawn out are not returned to the jar.

then two white balls in five successive draws without replacement is

$$\frac{30 \times 29 \times 28 \times 20 \times 29}{50 \times 49 \times 48 \times 47 \times 46}$$

The probability of drawing black, then white, then black, then white, then black in five successive draws without replacement is

$$\frac{30 \times 20 \times 29 \times 19 \times 28}{50 \times 49 \times 48 \times 47 \times 46}$$

The probability of drawing three black and two white balls in any one specified order in five draws without replacement is

$$\frac{30 \times 29 \times 28 \times 20 \times 19}{50 \times 49 \times 48 \times 47 \times 46}$$

What is the probability of drawing, in five successive draws, three black balls and two white balls, irrespective of the order in which they appear?

This one is not quite so easy. The various orders are mutually exclusive (if you get *one* order, you do *not* get any other order), so the probability in question is the sum of the probabilities of all the various possible orders. The probabilities of these various orders are all equal: so we only have to multiply the probability for some one specified order (which we just calculated) by the number of possible orders — that is, by the number of permutations of five things, three of which are alike of one kind, and two of which are also alike but of another kind (the two white balls).

This requires the formula for the number of permutations of objects some of which are alike. We wrote down the formula for this situation on page 96, but did not prove it there. It's so easy and such fun that we will do it now.

Take first the following case: how many permutations are there of n objects, a of which are alike, the remaining $n - a$ being distinct? Put some sort of identifying mark (spots of different colors, or whatever) on each of the a things which are alike so that you can now distinguish between them.

Suppose we call x (that mysterious, unknown, but useful x) the number of permutations of n things of which a are alike. Suppose one of these permutations — one of these "orders" to be spread out in front of you. We will call this "Order No. 1." We have x such orders, but as yet we don't know what x is.

If you interchange two of the alike objects, this continues to be the same permutation; for you can't tell that two, being alike, were interchanged. But if you have secretly identified the a otherwise identical objects, then by changing them into all possible orders (all possible orders of a objects) you would get, out of Order No. 1, as many permutations as there are ways of ordering a objects — which we know to be $a!$

Now take Order No. 2, and make $a!$ permutations out of it, by changing around the (temporarily identifiable) a objects it contains.

You can do this with every one of the x orders, and you thus make, in all, $x \times a!$ permutations. These are all *distinct*, since they differ either in the order of the non-alike objects, or in the order of the alike (but secretly labeled) objects. And clearly we have in this way manufactured all the orders one can get from n

objects, which are all distinct — which we know to be $n!$ in number. That is

$$x \times a! = n!$$

or

$$x = \frac{n!}{a!}$$

This is a very reasonable formula. It says that the number of permutations of n objects a of which are alike is, of course, smaller than the number, $n!$, which you could get if they were all distinct; and the number is cut down by the factor $a!$ which precisely measures the order variety which you can achieve with a objects if they are different, and which you *lose* because these a objects are alike.

You can see at once that the reasoning generalized without the slightest difficulty. If you have a alike and of one kind, b alike and of a second kind, c alike and of a third kind . . . then the permutation of n objects is

$$\frac{n!}{a!\, b!\, c! \cdots}$$

So now we can answer the question that started us off on this side issue. The probability of drawing, in five successive draws, three black and two white (but in any order) is

$$\frac{30 \cdot 29 \cdot 28 \cdot 20 \cdot 19}{50 \cdot 49 \cdot 48 \cdot 47 \cdot 46} \cdot \frac{5 \cdot 4 \cdot 3 \cdot 2 \cdot 1}{(3 \cdot 2 \cdot 1)(2 \cdot 1)}$$

The second term here has the value 10; so that the probability of your drawing 3 black and 2 white balls *in an unspecified order*, when 5 are drawn in succession without replacement, is 10 times as great as the probability that you will draw 3 black and 2 white in a specified order.

Now that you know how (and don't be fooled by the black and white balls; this is a very fundamental kind of problem), what is the probability of drawing, in nine successive draws, two black and seven white, regardless of order? Check that it is

$$\frac{30\cdot 29\cdot 20\cdot 19\cdot 18\cdot 17\cdot 16\cdot 15\cdot 14}{50\cdot 49\cdot 48\cdot 47\cdot 46\cdot 45\cdot 44\cdot 43\cdot 42} \text{ times}$$

$$\frac{9\cdot 8\cdot 7\cdot 6\cdot 5\cdot 4\cdot 3\cdot 2\cdot 1}{(2\cdot 1)(7\cdot 6\cdot 5\cdot 4\cdot 3\cdot 2\cdot 1)}$$

Note about Decimal Expansions

If you divide 2 by 3 to produce a decimal figure, you get 0.66666 · · · , the "6's" going on forever. If you divide 7 by 11 you get 0.63636363 · · · , the pair "63" repeating indefinitely. These are what are called "repeating decimals." If you divide any integer whatsoever by any other integer, leaving a remainder, you get a result the decimal part of which, after a time, repeats itself indefinitely. Thus

$$\frac{137}{242} = 0.566115702479338842975\,20\,661157024793\cdots$$

There are 35 rather wildly assorted digits — but there is a pattern which repeats itself indefinitely.

Any number which can be written as the quotient of two integers is called, in mathematics, a *rational number*. A number such as $\sqrt{2}$ can be *approximated* by the quotient of two integers. Thus $\sqrt{2}$ is roughly equal to 14/10, is more closely approximated by 141/100, and still more closely by 14142/10000. But none of these fractions is exact: and no such fraction *can* exactly equal $\sqrt{2}$.

If you solve a simple algebraic equation such as

$$x^2 - \frac{5x}{6} + \frac{1}{6} = 0$$

you get, as answers, $x = 1/2$ and $x = 1/3$. These are both *rational* numbers. Similarly there are equations which lead, as their solutions, to *irrational* numbers. But there are numbers which simply cannot turn up when you solve ordinary algebraic equations with rational numbers as coefficients of the unknown (as 1, $-5/6$, and 1/6 are coefficients in the equation just written). Such numbers are called by mathematicians *transcendental* numbers.

The famous numbers e and π are transcendental numbers. In the decimal expansion of either of them there is no "pattern" whatsoever for the way the digits appear. In the expansion of π, for example, the 501st to the 521st digits are

98336 73362 44065 66430

whereas the 1001st to the 1021st are

38095 25720 10654 85863

and the 2001st to the 2021st are

94657 64078 95126 94683

A study has been made (*Mathematical Tables and Other Aids to Computation*, Vol. IV, No. 29, pp. 109–11) of how *random* is the appearance of the various digits in these long decimal expansions. In the case of the expansion of e to 2000 digits the various digits appear as follows

digit	0	1	2	3	4	5	6	7	8	9
number of times digit appears	196	190	208	202	201	197	204	198	202	202

These values, interestingly, are rather *closer* to the average figure 200 than one would expect if they were distributed at random.

In the latest calculations[11] of e and π, using the most powerful computer, the first 5000 digits in the decimal expression for π were computed, on the machine, in less than one minute!

[11]The value of e was calculated in 1953 to 60,000 digits on the Illiac (D. J. Wheeler, Digital Computer Laboratory Internal Report No. 43, University of Illinois); whereas π has been computed, in 8 hrs. 43 min., on IBM 7090 to 100,000 digits (Daniel Shanks and John W. Wrench, Jr., *Mathematics of Computation*, Vol. 16, No. 77, January 1962).

Chapter VII

MATHEMATICAL EXPECTATION

Fate, Time, Occasion, Chance, and Change
— to these all things are subject.

PERCY BYSSHE SHELLEY

How Can I Measure My Hopes?

A jar contains 100 balls, 80 of which are "blanks" with no numbers at all. Of the remaining twenty balls, fifteen are numbered 2, four are numbered 10, and one is numbered 25. They are all thoroughly mixed up, and I am allowed (blindfolded) to draw out one. I will receive a prize, equal in dollars to the number on the ball I draw. What are my reasonable hopes of return?

I stand in front of a slot machine putting in a dime, or a quarter, or even a silver dollar if I am in one of the famous Western gambling towns. Maybe I will hit the jackpot. Probably I will get nothing. What can I reasonably hope — or, more accurately, *expect?*

These questions, and many more which are similar but more serious and important, illustrate the problem one faces when he is confronted with alternatives. Each alternative is worth a certain amount to the player — or would cost him a certain amount — and to each can reasonably be assigned a probability

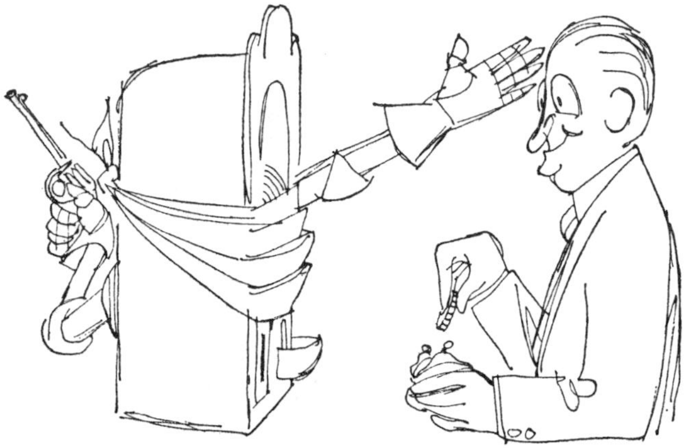

FIG. 16

that it will, in fact, turn out to be the alternative that occurs.

This sort of problem was considered fairly early after the birth of Lady Luck. In his "Essay on the Analysis of Games of Chance" (to translate the title into English), published in 1708, Montmort reported that Nicolas Bernoulli (1687–1759)[1] had written to him about the following problem:

B is rolling a die. A agrees to give B one crown if A gets a six on the first roll, two crowns if B gets his first six on the second roll, three crowns if B gets the first

[1]The Bernoulli family nurtured multitudinous mathematicians in the seventeenth and eighteenth centuries. There were eight of them: *Jacques*, 1654–1705, called *James* by English writers; his brother *Jean*, 1667–1748, called *John* by English writers; *Nicolas*, 1695–1726, eldest son of Jean; *Daniel*, 1700–1782, second son of Jean; *Jean*, 1710–1790, youngest son of Jean; *Nicolas*, 1687–1759, cousin of the preceding three; *Jean*, 1744–1807, son of Jean and grandson of Jean; and *Jacques*, 1759–1789, his younger brother.

six on the third roll, and so on. What can B expect to realize from this game?

Mathematical Expectation

In order to answer questions of this sort, the early masters introduced the idea of *mathematical expectation.* But let us start with a somewhat simpler example.

Suppose that I am playing a game by tossing a coin; and I get a prize of $10 if the coin comes up heads — for which the probability is, of course, 1/2 — but get nothing if the coin comes up tails — for which the probability is also 1/2. My *mathematical expectation* is obtained by multiplying the magnitude of the prize ($10) by the probability (1/2) of getting it. Thus my mathematical expectation in this example is $5. As a purely formal point, I ought to multiply the alternative prize by its probability, and add the product to the former one to produce the total mathematical expectation. But the alternative prize is *zero;* so it contributes nothing to the total expectation.

Note that what really happens, on any single trial of this game, is that I get nothing if the coin comes up tails, or $10 if the coin comes up heads. In no individual case can I actually get $5, which is the mathematical expectation.

But if we play this game a lot of times (say $2n$ times with n a large number), then I ought to win $10 on each of about half of the tosses (or on about n tosses), and I ought to win nothing on each of about the other half of the tosses (also about n tosses). So the *average amount* I could expect to win per toss would be the average of about n times \$10 plus about n times zero. This average[2] is

$$\frac{\overbrace{\$10 + \$10 + \cdots + \$10}^{\text{about } n \text{ terms}} + \overbrace{\$0 + \$0 + \cdots + \$0}^{\text{about } n \text{ terms}}}{2n}$$

which is about

$$\frac{n \cdot \$10}{2n} = \frac{\$10}{2} = \$5$$

That is to say, since the probability of an event is a good estimate of the fraction of times the event will occur in a long series of trials,[3] the *mathematical expectation* is a good estimate of the average return (or loss) one will have in a long series of trials.

In general, if we are thinking of trials which can come out in several ways that have rewards or prizes equal to V_1, V_2, V_3, ... (the letter V being chosen because it is the initial letter of *value*, some of these values being possibly negative), and if the probabilities of getting these various prizes of values V_1, V_2, etc., are P_1, P_2, etc., then the mathematical expectation in this "game" is defined to be

$$\text{Mathematical expectation} = P_1V_1 + P_2V_2 + P_3V_3 + \cdots \quad (13)$$

The Jar with 100 Balls

Let's use this definition to answer the questions with which this chapter opened. In this game there

[2]When you want the ordinary arithmetical average of several numbers, you simply add them together, and divide by the number of numbers.

[3]We will later have a more substantial basis for this remark, but at the moment it should at least sound sensible.

are four distinct things that can happen. I can draw a blank (for which the probability is 80/100), or I can draw a \$2 prize (for which the probability is 15/100), or I can draw a \$10 prize (for which the probability is 4/100), or I can draw a \$25 prize (for which the probability is 1/100). That is, there are four possibilities:

$$\begin{aligned} V_1 &= 0 & P_1 &= 80/100 \\ V_2 &= \$2 & P_2 &= 15/100 \\ V_3 &= \$10 & P_3 &= 4/100 \\ V_4 &= \$25 & P_4 &= 1/100 \end{aligned}$$

and the mathematical expectation is therefore

$$\$0 \cdot \frac{80}{100} + \$2 \cdot \frac{15}{100} + \$10 \cdot \frac{4}{100} + \$25 \cdot \frac{1}{100} = \frac{\$95}{100} = \$0.95$$

In a long series of plays I could reasonably expect to receive a return averaging about \$0.95 per game. A good deal of the time I would receive nothing (in fact, in about 80 per cent of the plays). In about 15 per cent of the plays I would receive \$2. In about 4 per cent I would receive \$10. In about 1 per cent I would receive \$25.

What would be a reasonable fee for playing this game? If we disregard any costs or profits for the "house," then clearly if one pays 95 cents per play he will, in the long run, just about come out even.[4] If the player can find any gambler who will let him play this game for less than 95 cents per trial, then

[4]This remark needs some corrective caution, which we can't very well make until later.

he has found a foolish gambler who will lose money. If the player pays more than 95 cents per trial, then this extra amount is, in the long run, profit for the gambler and loss for the player.

The player might say, "The gambler is entitled to a little edge, since gambling is his business and he has to eat"; and he might also say, "I enjoy the excitement of this game, and this excitement is worth something to me." These thoughts would justify the player in paying a little more than 95 cents per game.

Suppose the player has time to play only a few games, or suppose he has in his pocket only about $5. Then how should he think about this game?

The phrase, "in a long series of trials," does not apply to him if he has time to play only four games, say. Now notice the sources of the partial contributions to the total mathematical expectation of 95 cents. The player gets no return at all from what happens about 80 per cent of the time. He gets a $2 prize from what happens about 15 per cent of the time — and this contributes 30 cents to the over-all mathematical expectation. Similarly 40 cents of the mathematical expectation comes from a result which will happen only 4 per cent of the time, and 25 cents from a result that will happen on the average only about once in a hundred times.

If this busy player has time for only four games the probability is

$$\frac{80}{100}\cdot\frac{80}{100}\cdot\frac{80}{100}\cdot\frac{80}{100} = 0.4096$$

that he will draw four blanks in a row and get nothing! That is, the odds are about 4 to 6 that he will receive no prize at all, and just throw away his investment of 4 × 95 cents or $3.80.

On the other hand, what is the probability that he will draw three blanks and one $2 prize?

The probability would be

$$\frac{15}{100}\cdot\frac{80}{100}\cdot\frac{80}{100}\cdot\frac{80}{100} = 0.0768$$

that he would start out with the $2 prize and then draw three blanks. But he might get the prize on the second, or third, or fourth trial; so the total probability (these orders are mutually exclusive) of three blanks and a $2 prize in any order is[5]

$$4 \times 0.0768 = 0.3072$$

so this hurried player has a probability of about 0.31 (or the odds are about 31 to 69) that he will invest $3.80 and get back $2. Combining this with the previous calculation, and noting that the events are mutually exclusive so that the probabilities add, the probability is 0.41 + 0.31 = 0.72 (or the odds are about 72 to 28) that the four games will leave him with a loss.

You can easily go on and calculate the probability that he will get three blanks plus one $5 prize (which would give him a modest profit of $1.20). He *could*, of course, draw the $25 ball four times in a row! That would give him a handsome profit of $96.20, but the probability is

$$\left(\frac{1}{100}\right)^4 = 10^{-8}$$

or one in one hundred million.

[5]Why the factor 4?

Somewhat similar, but more complicated, considerations enter if our player has only a few dollars in his pocket. For then, if he is prudent, he must face the fact that he may have a run of "bad luck" which will wipe out his resources and force him to quit so that he has no opportunity to experience what should happen "in the long run." If he has only $5 he can, at the beginning of the game, stand only four successive blanks — and we have seen that four such successive blanks must be expected about 41 per cent of the time. If he makes a winning draw before the fifth game, then his resources are thereby augmented. If this is a $2 draw, he is still in a rather shaky position. If he is very lucky and makes the $25 haul before the fifth game, then his resources have improved so much that he can reasonably expect to play for quite a time.

There is a large body of results in modern probability theory concerning "runs," the chances of long uninterrupted series of results of one kind, etc. We will have a little to say about this subject later, but in general the topic is a bit too complicated for this book. You can see, even from the present trivial example, that a knowledge of the theory of runs can be of very practical value.

The One-Armed Bandit

Let's look now at the second question with which this chapter opened. My friend Professor Philip G. Fox, who teaches statistics to students in the School of Commerce at the University of Wisconsin, has for years used examples from horse racing, football pools, slot machines, and similar supposedly easy-money gimmicks, to enliven his teaching of the basic facts of probability and statistics.[6]

A typical slot machine has three dials, each with 20 symbols (cherries, oranges, lemons, plums, bells, and bars). Since each of the three dials may come to rest in any one of 20 positions, there are $20 \times 20 \times 20 = 8000$ different permutations possible. A typical set of dials is arranged like this:

TABLE VI
DIALS ON A TYPICAL SLOT MACHINE

	Number on		
	Dial I	*Dial II*	*Dial III*
Cherries	7	7	0
Oranges	3	6	7
Lemons	3	0	4
Plums	5	1	5
Bells	1	3	3
Bars	1	3	1
	20	20	20

meaning, for instance, that Dial I has 7 cherries on it, Dial II has 7 cherries, Dial III has no cherries, Dial II has 1 plum, etc.

The machine pays off in accordance with the following table:

[6]See "Primer for Chumps," by Professor Philip G. Fox, *The Saturday Evening Post*, November 21, 1959.

TABLE VII
PAYOFF ON TYPICAL SLOT MACHINE

Paying Permutation			*Payoff*	*Average Number of Occurrences per 8000 Trials*	*Average Total Paid Out per 8000 Trials*
Bar	Bar (Jackpot)	Bar	62	3	186
Bell	Bell	Bell	18	9	162
Bell	Bell	Bar	18	3	54
Plum	Plum	Plum	14	25	350
Plum	Plum	Bar	14	5	70
Orange	Orange	Orange	10	126	1260
Orange	Orange	Bar	10	18	180
Cherry	Cherry	Lemon	5	196	980
Cherry	Cherry	Bell	5	147	735
Cherry	Cherry	X[7]	3	637	1911
Average number of winning plays in 8000 trials				1169	5888
Average number of non-winning plays in 8000 trials				6831	

You can see what a clever arrangement this is. Dial I comes to rest first; and there are only 3 chances out of 20 that it shows a lemon so that your hopes immediately vanish. There are 5 chances out of 20 that Dial I will show a plum. But with a plum on Dial I you have to get a plum on Dial II to stay in business, and the chance of that is only 1 in 20. And if you do experience the excitement of a plum on Dial II, you have to get, on Dial III, either a

[7]X can be anything except a Lemon or Bell.

plum or a bar — or again you are out of business. The chance of a plum or a bar on Dial III (total probability of two mutually exclusive events) is 6/20.

The "payoff" for the various winning permutations is, of course, built into the machine. The average number of occurrences of any one permutation, such as "Plum, Plum, Bar," for example, can be calculated from the fact that Dial I can produce Plum in 5 ways (see the table above); Dial II, Plum in 1 way; and Dial III, Bar in 1 way. Hence Plum, Plum, Bar can be produced in $5 \cdot 1 \cdot 1 = 5$ of the 8000 ways.

You can see from the last table, which I understand is a typical one for a "commercial" machine (some private clubs have machines which pay better), that in a long run of trials with a dollar machine, you could expect about $5888 return for every $8000 "invested." That is to say, you could expect to lose roughly one-quarter of the money you put into the slot.

Professor Fox has calculated, for the dial settings given above, the following interesting and presumably sobering data:

Probability of a win on any one play	= 0.146
Probability of a run of two wins	= 0.021
Probability of a run of three wins	= 1/320
Probability of a run of four wins	= 1/2193

The probabilities of a series of "blank" runs with no win at all are:

TABLE VIII

PROBABILITY OF SUCCESSIVE FAILURES WITH SLOT MACHINES

Number of Blank Trials	*Probability*
1	0.854
2	.729
6	.388
10	.206
20	.042
40	.002

And, perhaps the most interesting and impressive of all:

TABLE IX
PROBABILITY OF WINNING AT SLOT MACHINE

Probability that after *n* plays
one will still be a winner

n	*Probability*
1	0.375
10	.127
20	.085
50	.026
100	.004
200	.00012
300	.00000

So if you are perseverant enough at this game, you can confidently expect to end up in the hole. With a 25 cent machine, and playing 12 times per minute, you can expect to lose just about $45 per hour.

All these results you should by this stage be able to calculate for yourself if you are interested, although some of the calculations would be rather tedious.

In a later chapter, and in a more general connection, we will make some comments about "systems" of play. At the moment merely accept the bald statement that there is no way to beat the probabilities listed.

The Nicolas Bernoulli Problem

Let's look now at the Nicolas Bernoulli question, mentioned near the beginning of this chapter. Remember that A (the "house") gives B (the player) one crown if he rolls a six as the first toss of a die, two crowns if he first gets a six on the second toss, three crowns if he first gets a six on the third toss, and so on.

Although it is *exceedingly* likely that a six will appear before, say, the fiftieth toss, it is theoretically possible that this would not occur. In fact, there is in theory an infinite number of possible outcomes of this game, accordingly as the first six appears on toss number 1, or 2, or 3, or 4 . . . or 967, or 968 . . . and so on forever. This is a case in which Equation (13), which defines mathematical expectation, has an indefinitely large number of terms on the right-hand side.

The probability P_1 of getting the prize $V_1 = 1$ crown by rolling a six on the first trial is 1/6. The probability for $V_2 = 2$ crowns is $5/6 \cdot 1/6$ since this is the compound probability of the independent events "not getting a six on the first throw" and "getting a six on the second." Similarly for $V_3 = 3$ crowns the probability $P_3 = (5/6)^2 \cdot 1/6$. The general formula for the probability of winning the n'th prize $V_n = n$ crowns is

$$P_n = (5/6)^{n-1} \times 1/6$$

since V_n is won only if you fail to get a six $n - 1$ times in a row and then do get a six.

Thus the mathematical expectation is, for this game

$$\frac{1}{6}\,(1\text{ crown}) + \frac{5}{6} \times \frac{1}{6}\,(2\text{ crowns}) + \left(\frac{5}{6}\right)^2 \times \frac{1}{6}\,(3\text{ crowns}) + \cdots$$
$$\cdots + \left(\frac{5}{6}\right)^{n-1} \times \frac{1}{6}\,(n\text{ crowns}) + \cdots$$

This is what mathematicians call an *infinite series*, since it goes on forever. Sometimes such series are *convergent*. That is, sometimes, as you add up more and more terms, you keep getting numbers which are closer and closer to some perfectly good finite number. For example, if you look at the infinite series

$$1 + \frac{1}{2} + \frac{1}{4} + \frac{1}{8} + \frac{1}{16} + \frac{1}{32} + \frac{1}{64} + \cdots$$

you can easily see that the sum of more and more terms keeps on getting closer and closer to 2, which is therefore called the sum of the series. Of course many infinite series do not behave in this way, but *diverge* in the sense that they keep on getting larger and larger, as does

$$\frac{1}{2} + \frac{1}{2} + \frac{1}{2} + \frac{1}{2} + \frac{1}{2} + \cdots$$

or jump around erratically, as does

$$1 - 1 + 1 - 1 + 1 - 1 + 1 \cdots$$

The series written down for the mathematical expectation in the Bernoulli game is a convergent one, and its sum is 6 crowns.

So if B plays this game a large number of times, he

can expect to average about 6 crowns a game of prize money. If A charges (as any gambler who expects to stay in business must) more than six crowns per game to play, then B has to charge off (the excess over six crowns) as worth while from the point of view of excitement, or entertainment, or the inevitable cost of growing up.

The St. Petersburg Paradox

Shortly after Nicolas Bernoulli invented the game we have just discussed, debate began concerning another game, somewhat similar but much more interesting.

In this game, B, the player, tosses a coin; and A, the "house," agrees to pay \$2 to B (we shift to the modern unit of dollars to be able to visualize the amounts more easily) if B gets a head on the first toss, \$4 if B first gets a head on the second toss — and so on, doubling the prize every time the winning head is delayed one more toss.

What is B's mathematical expectation in this game? The chance of a head on the first toss is 1/2. The chance of a tail on the first toss and a head on the second toss is $1/2 \cdot 1/2$. The chance of a tail on each of the first $n - 1$ tosses and then a head on the n'th toss is

$$\left(\frac{1}{2}\right)^{n-1} \cdot \frac{1}{2} = \frac{1}{2^n}$$

The prize if the head is delayed until the n'th toss is

$$\$2^n$$

Hence the mathematical expectation of B is

$$\frac{1}{2}\cdot\$2 + \frac{1}{4}\cdot\$4 + \frac{1}{8}\cdot\$8 + \frac{1}{16}\cdot\$16 + \cdots + \frac{1}{2^n}\cdot\$2^n + \cdots$$

this series going on indefinitely. But each term of this series, if we multiply out, is \$1. Thus the mathematical expectation of B is

$$\$1 + \$1 + \$1 + \cdots$$

Now this, if you think a bit, seems almost ridiculous. Archimedes proved that, by adding one to itself a sufficient number of times, you can produce a sum which exceeds any number you care to name, however large. This fact is, to be sure, self-evident. It says simply that the sequence of positive integers

$$1, 2, 3, 4, 5, 6, 7 \ldots$$

never reaches a point where it gets stuck and can go no further.

But this, in turn, says that B's expectation in this game is "infinite"; for that often abused word means, precisely and simply in this connection, that the sum of the series which represents the expectation can be made to exceed any number, however large, just by taking enough terms.

No matter what B offers to pay A, in order to play the game, A says, "Not enough."

But is it really worth an indefinitely large amount to play this game? Of course it is not! It would, for most persons at least, be very silly to pay such a sum as \$1 million, which is a negligibly small fraction of what in theory the game is worth.

Think about it this way. If I am offered a 50–50 chance to win \$2, that chance is reasonably worth \$1. I can afford to bet \$1, for it will not ruin me if I lose. And if I win, I am confident that the house will pay off.

But in the St. Petersburg game, neither of these remarks applies. Although \$1 million is, from the point of view of the formal theory, a very cheap entrance ticket, it is an impossible price for me, partly because I just haven't that kind of money, and partly because it doubtless would ruin me to lose that amount, even if I had it. Second, the so-called "infinite value" of the St. Petersburg game depends essentially upon the house's being able to pay off, no matter what happens.

It is very illuminating to see what this game is worth to *B*, taking into account the capacity of *A* to pay off.

Fig. 17

If *A* hasn't got an "infinite" amount of money with which to pay off (and of course he *can't* have), the series which represents the mathematical expectation in this game changes when the payoff exceeds A's capacity to pay. If, in fact, A has 2^m dollars rather than impossible "infinite" resources, then the expression for the mathematical expectation of B suddenly changes at the point where A is supposed to pay *more* than 2^m dollars. From that point on, the most A can do is to pay all he has — namely, 2^m dollars.

Thus under this more realistic assumption about the banker, the mathematical expectation of the player is

$$\frac{1}{2}\cdot \$2 + \frac{1}{4}\cdot \$4 + \cdots + \frac{1}{2^m}\,\$2^m + \$2^m\left[\frac{1}{2^{m+1}} + \frac{1}{2^{m+2}} + \frac{1}{2^{m+3}} + \cdots\right]$$

$$= \underbrace{\$1 + \$1 + \cdots + \$1}_{m \text{ terms}} + \$1\left[\frac{1}{2} + \frac{1}{4} + \frac{1}{8} + \cdots\right]$$

$$= m + 1 \text{ dollars}$$

On this basis, the game is worth not an "infinite" amount, but $m + 1$ dollars. Let's look at a little table:

Value of m	2^m*Dollars: that is to say the Resources in Dollars, of the "House," A*	*Value of Game to the Player, B, in Dollars*
0	1	1
1	2	2
2	4	3
4	16	5
6	64	7
8	256	9
16	65.536	17
32	4.3 billion	33

In other words, and strictly according to the theory, mind you, it is worth only the moderate amount of \$9 to play this with one of your friends who is going to faint, or pretend it was all a joke, or drop dead, if he has to pay more than \$256; it is worth \$17 when played against a professional adversary who really will pay off up to about \$65,000; and is worth \$33 when played against a syndicate which can and will pay off up to \$4.3 billion.

This makes a great deal of sense — as probability theory does if you treat it right.

It is illuminating also to see what happens to this game if, instead of paying off in multiples of \$2, it pays off in multiples of an only slightly smaller amount. To illustrate, suppose A, the house, agrees to pay B, the player, \$1.95 if B gets a head on the first toss, $(\$1.95)^2$ if B gets the first head on the second toss, $(\$1.95)^3$ if B gets the first head on the third toss, and so on.

This, at first sight, does not seem to differ very greatly from the original form of the game. B might easily think, "I get almost as much as in the original game," and A might think, "I have to pay almost as much."

But this is only because they would be thinking of B's getting a head on a fairly early toss. If the win is long delayed, this second game differs more and more drastically from the first. The ratio of $(\$1.95)^n$ to $(\$2.00)^n$ is, of course, $(0.975)^n$: and this ratio gets smaller and smaller as n increases. Thus this second game does not involve the extremely large (even though very unlikely) prizes that made the first game impossibly expensive.

The expectation of B for this second game, in fact, is

$$\frac{1}{2}(\$1.95) + \frac{1}{2^2}(\$1.95)^2 + \cdots + \frac{1}{2^n}(\$1.95)^n + \cdots$$

and the sum of this series is not infinite, but works out[8] to be exactly \$40.

Thus B ought to pay A only \$40 a time to play this game. Can you now figure out that if B is willing to pay only \$10 a game, B ought to offer to play on the basis of prizes which are multiples not of \$2 or of \$1.95, but of \$1.80?

Daniel Bernoulli, about 1720, tried to improve the concept of mathematical expectation by making a distinction between what he called *physical fortune* and *moral fortune*. The physical fortune is the actual amount of money you have; but the moral fortune involves the concept of how important additional

[8] This will be at least plausible if you divide 1 by $1 - x$ by ordinary long division (ordinary except that letters are involved rather than just numbers); you get

$$\begin{array}{r|l}
1 - x & 1 \qquad\qquad\qquad\qquad \big| \, 1 + x + x^2 + x^3 + \cdots \\
\hline
& \underline{1 - x} \\
& \quad x \\
& \quad \underline{x - x^2} \\
& \qquad x^2 \\
& \qquad \underline{x^2 - x^3} \\
& \qquad\quad x^3 \\
& \qquad\quad \underline{x^3 - x^4} \\
& \qquad\qquad x^4
\end{array}$$

It is in fact true that, for values of x less than 1, the series $1 + x + x^2 + x^3 + \cdots$ has a sum which converges to the value $1/(1 - x)$. If you substitute the value 1.95/2.00 for x, the sum comes out to have the value 40.

money is to you in view of what you have already.

He said in effect: If I add a dollar to my wealth this means a great deal to me if I do not have much money; but it means less and less to me the more I already have. Thus he started out with assuming that

$$\text{increase in } \textit{moral fortune} = \frac{(\text{constant})(\text{increase in } \textit{physical fortune})}{(\text{size of physical fortune})}$$

Using this idea, B's mathematical expectation in the St. Petersburg game is finite, even though one assumes that the house can pay off, no matter how much B wins. This idea of moral expectation has a certain historical and curious interest, but it never proved to be of practical value.

Summary Remarks About Mathematical Expectation

If one is dealing with games paying off in money, then the mathematical expectation measures, as we have seen, the reasonably expectable financial return per game. If the cost of playing the game is included in the calculation, then for a *fair game* one would have the equation

Total Mathematical Expectation in Fair Game =

Sum of [(probability of winning each prize)(value of each prize)] − Cost of Playing = 0

the value of zero resulting from the fact that costs, losses, and wins should just balance out if the game is really fair.

In a more general situation, if a quantity x (and this *may* be the money prize in a game, or the result of carrying out some measurement or experiment of any sort, or the estimated value of some procedure) can take on the n discrete values $x_1, x_2, x_3 \ldots x_n$, and if the probabilities that x does assume these values are $P(x_1), P(x_2), \ldots P(x_n)$, then the *mathematical expectation in x*, or the *expected value of x* (these two phrases mean exactly the same thing), denoted by the symbol $E(x)$ is defined as

$$E(x) = P(x_1)\cdot x_1 + P(x_2)\cdot x_2 + \cdots + P(x_n)\cdot x_n \quad (14)$$

Further, since $P(x_1)$ is a reasonable measure of the fraction of the cases in which x comes out to have the value x_1 (and the similar remark for all the other possible values of x), this expectation in x is a reasonable measure of the *average value* that we can expect x to have in a long run of trials.[9]

To illustrate the use and meaning of formula (14), consider a box which contains a large number of cards, say 1000, which are identical except for the fact that 500 are marked 1, 300 are marked 2, and 200 are marked 3. We stir up the cards thoroughly and we are about to draw one card. What is our mathematical expectation of the number on the card?

It is

$$\frac{500}{1000} \times 1 + \frac{300}{1000} \times 2 + \frac{200}{1000} \times 3 = 1.7$$

for, all the cards being equally probable, the probability of drawing a card numbered 1 is 500 (the

[9]This remark, which we accept here on the intuitive basis of our present understanding of the relationship between *probability* and *frequency of occurrence*, will be better substantiated by the results discussed in the next chapters.

outcomes leading to the event 1) divided by 1000 (the total number of equally probable outcomes), etc.

Do we expect (using the word in its ordinary sense) that the card we will draw will have the number 1.7 on it? Of course that is ridiculous, for there is no card so numbered.

But suppose we drew a considerable number of cards (each time returning the card and restirring). Suppose, in fact, that we thus draw n_1 cards which are numbered 1, n_2 which are numbered 2, and n_3 which are numbered 3. What would be the average size of the numbers on all these $n_1 + n_2 + n_3$ cards? To get the ordinary numerical average of a set of numbers we remember that we simply add them and divide by the number of them. Thus

$$\text{Average number} = \frac{\overbrace{1 + 1 + \cdots + 1}^{n_1 \text{ terms}}}{n_1 + n_2 + n_3} + \frac{\overbrace{2 + 2 + \cdots + 2}^{n_2 \text{ terms}}}{n_1 + n_2 + n_3} + \frac{\overbrace{3 + 3 + \cdots + 3}^{n_3 \text{ terms}}}{n_1 + n_2 + n_3}$$

$$= \frac{n_1 \cdot 1 + n_2 \cdot 2 + n_3 \cdot 3}{n_1 + n_2 + n_3}$$

$$= \frac{n_1}{n_1 + n_2 + n_3} \cdot 1 + \frac{n_2}{n_1 + n_2 + n_3} \cdot 2 + \frac{n_3}{n_1 + n_2 + n_3} \cdot 3$$

Unless we have actually carried out the experiment we don't know what n_1 and n_2 and n_3 are. But

$$\frac{n_1}{n_1 + n_2 + n_3}$$

is the proportion, in our sample of $n_1 + n_2 + n_3$ drawn cards, with which cards were drawn bearing the number 1. If we drew a good-size sample, we could reasonably expect that the proportion of 1's in our sample would be close to 500/1000, which is the proportion of 1's in the box. Similarly we could expect $n_2/(n_1 + n_2 + n_3)$ to be close to 300/1000, and $n_3/(n_1 + n_2 + n_3)$ to be close to 200/1000.

Hence we can reasonably expect that the average of the numbers on the $n_1 + n_2 + n_3$ cards drawn would be

$$\text{reasonably expectable average} = \frac{500}{1000}\cdot 1 + \frac{300}{1000}\cdot 2 + \frac{200}{1000}\cdot 3 = 1.7$$

So the *mathematical expectation* of the number on one card to be drawn has a value which can reasonably be expected to be close to the actual average value if a lot of cards are drawn.

Using the various rules we discovered in Chapter V, it is not difficult to show that, when dealing with two different quantities x and y, the expectation in their sum $x + y$ is the sum of the expectation $E(x)$ and $E(y)$, whether or not x and y are independent. The expectation $E(x\cdot y)$ of their product is equal to the product $E(x)\cdot E(y)$, but only if x and y are independent. Independent, in this connection, means that the probability of a certain value for x is the same, no matter what value y has (and vice versa). That is, the value of one of the quantities does not affect the probabilities for the various values of the other.

Try These

(a) On four of ten cards is written \$1, on three is written \$2, on two is written \$3, and on one is written \$4. If, on drawing a card after the ten are shuffled, you get what the card says, what is it worth to play this game once?

(b) In the previous game, suppose a card is drawn but not exposed. The player has to guess the class of the card: i.e., is it a \$1 card, a \$2 card, or what? If he guesses the class correctly, he gets the prize. Otherwise he gets nothing. Now what is his expectation?

This cannot be solved until one has some information as to the probability that he will guess correctly, and this in turn depends on what system of guessing he uses.

If he is just as likely to guess one class as another, then his expectation is

$$\frac{1}{4}\cdot\frac{4}{10}\cdot\$1 + \frac{1}{4}\cdot\frac{3}{10}\cdot\$2 + \frac{1}{4}\cdot\frac{2}{10}\cdot\$3 + \frac{1}{4}\cdot\frac{1}{10}\cdot\$4 = \$0.50$$

for on this assumption, the probability that he will guess any class is 1/4, regardless of what class this is.

But if he distributes his guesses proportionally to the total stakes of the various classes, he chooses probabilities out of this table

Class	\$1	\$2	\$3	\$4
Total Stake	\$4	\$6	\$6	\$4
Probability	0.2	0.3	0.3	0.2

and his expectation (work it out) is \$0.52. By guessing class in this way he has improved his expectation.

If he distributes his guesses proportionally to the number of cards in the classes (work out the probabilities as in the preceding table) then his expectation turns out, as in the first case, to be \$0.50. Thus by being smart about this game, he can average 2 cents more per game. Is there any better way for him to guess class?

(c) Think of a set of trials of some event, each trial taking S seconds. Suppose that, on each trial, the probability of a success is p, and of a failure is $1 - p = q$. What is the expectation of the length of time that successes will persist?

First of all, you must have one assured success in order to start a succession of successes. Second, note that a duration of say five S is also a duration of two S, three S, and four S. So it is not right to multiply every duration ($2S$, $3S$, $4S$, $5S$ . . .) by the probability that it will occur and add.

It is easy if you deal only with increments to the duration. You are assured of a starting duration of S seconds. What is the probability of adding an additional (second) duration of S more seconds? Clearly, p. Then what is the probability that, after this success, you will add another additional duration of S? That can occur only if a success is followed by a success — in other words, the probability is p^2. Following up this lead, you see that the expected duration of success is

$$1 \cdot S + p \cdot S + p^2 \cdot S + p^3 \cdot S + p^4 \cdot S + \cdots$$
$$= S[1 + p + p^2 + p^3 + \cdots]$$

Here is another of those infinite series. This one is convergent and its sum is $1/(1 - p)$. Perhaps you will find this more plausible if you actually divide 1 by $1 - p$. Thus

$$
\begin{array}{l|l|l}
1 - p & 1 & 1 + p + p^2 + p^3 + \cdots \\
 & \underline{1 - p} & \\
 & \quad p & \\
 & \quad \underline{p - p^2} & \\
 & \qquad p^2 & \\
 & \qquad \underline{p^2 - p^3} & \\
 & \qquad\quad p^3 & \\
 & \qquad\quad \underline{p^3 - p^4} & \\
 & \qquad\qquad p^4 &
\end{array}
$$

And since $1 - p = q$, we can rewrite this very interesting expression for the expected duration of successes in trials of an event in which the probability for success on each trial[10] is p. In fact,

$$\text{expected duration of success} = \frac{S}{q}$$

The greater the probability p of success, the less the probability q of failure; and accordingly the greater the expected duration of success. Very reasonable, to be sure; but you can hardly claim that this particular, very simple and elegant formula, was intuitively obvious.

It shouldn't strain you very much to prove that the expected duration of failures is S/p. In fact, can't you argue this just by means of a little semantic trick?

[10]This means, of course, that the trials are independent, in the sense that what happens on any one trial does not depend on what happened on previous trials. There is a large and rich theory for trials in which the probability on any one trial *does* depend on what has happened previously.

Where Do We Eat?

Referring to a well-known national chain of roadside eating places, my wife says, "I only score them about *B* — say 85 on a scale of 100 — but they are always the same, and I am *sure* that any one I pick will turn out to deserve a score of 85."

If we pick places to eat we don't know about, there is about one chance in ten that we might pick an *excellent* place (score 96), one chance in ten that we will pick a *terrible* place (score 20), two chances in ten that it will be *fair* (score 75), two chances that it will be *very good* (score 90), and the remaining four chances that it will be *just good* (score 85).

Where do we eat? That is, how do the expected scores compare, for the known chain, and the unknown places?

Chapter VIII

THE LAW OF AVERAGES

The most important questions of life are, for the most part, really only problems of probability.

LAPLACE,
Théorie Analytique des Probabilités

The Long Run

A great many of the things that happen in this world appear to occur very erratically if one looks only at a few instances. Over a very wide range, however, these erratic phenomena "smooth out" and show a more and more regular behavior as the number of instances of occurrence, or of trial, is increased. You can furnish all sorts of examples of this from your own immediate experience. If you make a record of the occurrence of heads and tails in tossing a coin, you are bound to have all one or all the other after a single toss. After five tosses there may well be three of one kind and two of the other, and you will not be particularly surprised if there are four of one kind and only one of the other. But after 100 tosses you will expect to have not too far from half of each. You would be pretty surprised to have a four-to-one mixture of heads to tails (i.e., 80 heads and 20 tails) after 100 tosses. And if you had 80 per cent heads and 20 per cent tails after 1000 tosses you would justifiably examine the coin very carefully; on the average, only once in about a million billion billion billion trials, each of 1000 tosses, would you get a discrepancy — from an even break between heads and tails — as large or larger[1] than 800 to 200.

It would be interesting but trivial if this sort of behavior were exhibited only by coins and dice. The fact is, however, that births, deaths, and a very large number of the important human experiences between birth and death, when measured, or assigned numerical value or rank, or tabulated, exhibit behavior most usefully illuminated by the theorems which probability furnishes.

Since I appeal to your intuition and common sense in considering the way coin-tossing proceeds, perhaps you are thinking that intuition and common sense are enough, that you really could "guess" a sensible answer to any problem of this sort. This happens to be spectacularly not the case, as you will see in this and the following two chapters. Probability theory not only *refines* and makes *quantitatively precise* a good many ideas which *do* seem intuitively reasonable; it also reveals many things that intuitively seem very astonishing indeed.

In fact, people often use the phrase "the law of averages" to justify incorrect remarks. I will illustrate with two examples, one from baseball and one from coin-tossing. The illustrations may, of themselves, seem trivial. But the considerations in-

[1]Notice the "as large or larger." The chance of getting precisely 800 heads and 200 tails in 1000 tosses is *smaller* than indicated above.

volved are important. And these two examples rather well and simply show up some points about probability that are frequently misunderstood.

For the first illustration, note that Norm Cash, the Detroit Tigers' first baseman, led both leagues last year (1961) in batting, with a "percentage" (as they perversely say in baseball) of .361. Suppose at some time, well along in the season so that his fine batting record was clear, he failed to get a hit in eight successive times at bat. Some sportswriter is almost sure to say that "according to the law of averages," Cash is bound to get a hit.

There are, as far as I can see, only two reasonable ways of considering this situation; and the sportswriter is dead wrong whichever interpretation one adopts.

The first way results from being convinced that it is not possible to invent an idealized probability model which will apply usefully to this situation. In other words, the first viewpoint is simply that probability considerations do not apply. And in that case, of course, no one has any business at all to invoke "the law of averages."

The reasons for thinking it impossible to design or invent a useful probability model are, among others, the following: In a simple probability situation (one to which, incidentally, the law of averages *would* apply) the probability of the outcome in question (in this case, a hit) remains constant from trial to trial,[2] and the trials are strictly independent, one of an-

[2]Probability theory can, of course, deal with situations in which the trials are not independent and the probability changes from trial to trial, but the ordinary "law of averages" does not then apply.

other. But the chance of a hit on a given time at bat clearly depends on the physical condition of the batter, on whether or not he is overconcerned and will therefore "tighten up," on whether he is relaxed and confident, on whether a hit at this particular time is critically important, on whether the manager of the opposing team has ordered the pitcher to give the batter an intentional pass, etc., etc. And one has a strong feeling that what is involved here, as a critical element, is *skill* and *strategy* rather than *chance.*

One cannot be too promptly confident that reasons of this sort rule out probabilistic behavior; events which are determined by a large number of contributing causes often do show chance behavior. Indeed, no one doubts that the tossing of a coin is a fully determined phenomenon. But the exact position in the hand, the exact force with which it is sent upward,

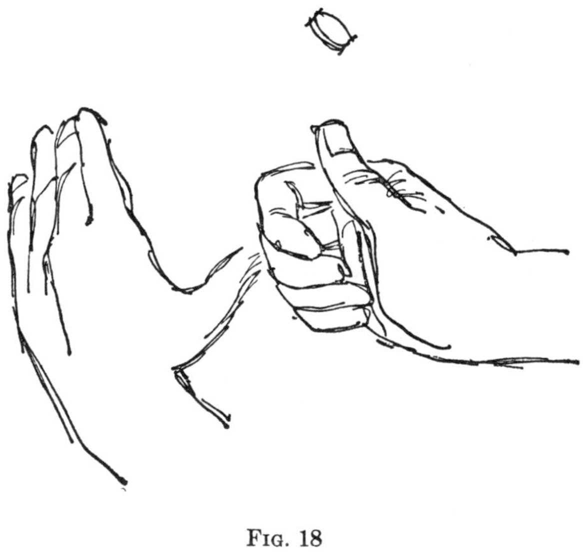

FIG. 18

the exact spin given to it, the exact distance down to the surface on which it falls, the exact elastic properties of that surface, and perhaps even the exact currents of air to which the coin is exposed — all these are so complicated, so intimately interrelated, and so practically inaccessible of measurement that the net result is a chance fall.

There just could be so many complicated and unpredictable factors involved in batting that "hits" actually would occur in the manner predicted by some mathematical probability model. But this is seriously debatable, and in our first view of this situation we are taking the position that probability considerations — at least of the simple sort justifying the phrase "the law of averages" — just do not apply. Adopting this view, the sportswriter was wrong.

But we can turn to the other view, and say that we believe, or are prepared to assume, that probability *does* apply and in the simple sense that the probability of a hit by Norm Cash on any time at bat is a constant, namely 361/1000. Then what about the sportswriter's claim?

We now are dealing with repeated and independent trials of an event which, on each trial, has the same probability for the outcome we have in mind. It is more convenient, and probably easier to avoid extraneous notions, if we do not talk about hits at repeated times at bat, but shift over to the familiar case of tossing a coin. Under the assumptions we have made the question of getting or not getting a hit is precisely like the question of getting or not getting a head, except for the numerical distinction of a probability equal to .361 in one case, and equal to .500 in the other.

Suppose, using a coin which you have carefully examined and balanced, you toss just as fairly as you are able, and you get eight heads in a row. Do you think that the "law of averages" tells you that, in these circumstances, there is better than an even chance for a tail on the ninth toss?

If you do, you are mistaken. The coin has no remembrance of the past. Each single time it is tossed the probability for a head is 1/2, and for a tail is 1/2. With a fair coin and with honest tossing, these values never change.

If a long set of tosses of a coin did start out with eight heads in a row,[3] then as the tossing proceeded this temporary peculiarity would gradually be *swamped* (as Professor Will Feller neatly puts it). It would be swamped partially by the time there had been 100 tosses, would be swamped largely after 1000 tosses, and would be swamped totally after 1,000,000 tosses. Being progressively swamped, as the tossing proceeds, does not in the least require that a toss which follows a run of heads should have any special responsibility to compensate for these past occurrences.

As is most usefully pointed out in the Mosteller-Rourke-Thomas book to which I have referred before, there are cases in which the probability of the occurrence of a certain event *does* increase as it is longer and longer postponed. If, for example, a box contains 100 balls, with 99 white balls and 1 black ball, and if one draws *without returning them* one

[3]This is, of course, possible. The probability of its occurring is $(1/2)^8$ or 1/256. In a very long series of trials of coin-tossing, about one of every 256 should start out with 8 uninterrupted heads.

white ball after another, then the probability of drawing the black ball gradually and steadily increases. The probability of drawing the black ball was only 1/100 on the first draw; but if one has drawn 90 white without returning any of them, then the probability of drawing the black ball has materially increased to the value of 1/10.

But a baseball player who is going to end the season with a batting average of .361 and who is going to come up to bat 1000 times (we assume this latter figure just to make the number simple) does not start out the season with a fixed future supply of 361 hits and 639 hitless times at bat and with the comforting assurance that as he "uses up" several hitless innings, the remaining proportion of hits will become larger.

Heads or Tails

Let us return again to the case of the tossed coin, and discuss this point a little more, for it is really a hard one for many persons to grasp or accept. When I asked my wife what she would expect to happen next if a fair coin, fairly tossed, was known to have come up heads eight times in a row, she replied, "It just stands to reason that you are much more liable to get a tail on the ninth toss — for everyone knows that in the long run the heads and tails balance out, and some balancing is obviously overdue."

Plausible as is this remark and apparently paradoxical as is the remark that even though heads and tails *will* tend to balance out in the long run, the probability of a head on the ninth toss is nevertheless exactly 1/2. The plausible remark is wrong and the apparently paradoxical one is correct.

At least part of the difficulty here results from the fact that many persons without much numerical experience in such matters completely fail to distinguish between heads and tails tending to balance *in the ratio sense,* and heads and tails tending to balance (or not balance) *in an absolute sense.* Let us look at this point a little.

Here is a table[4] for the way a coin-tossing experiment might proceed.

TABLE X
POSSIBLE RESULTS IN A COIN-TOSSING EXPERIMENT

No. of Tosses	*No. of Heads*	*No. of Tails*	*Ratio of Heads to Total Tosses*	*Absolute[5] Excess of Heads over Tails*
100	54	46	0.540	8
500	254	246	0.508	8
1000	501	499	0.510	2
5000	2516	2484	0.503	32
10,000	4979	5021	0.498	42

Notice in the fourth column that, as the number of tosses increases, the ratio of heads to total tosses starts out with the value 0.540, which is 8 per cent away from 1/2; and ends up 0.498, which is only 0.4 per cent away from 1/2. That is to say, this ratio

[4]These figures simulate a coin-tossing experiment, but are in fact a record of the occurrence of even digits in a section of *A Million Random Digits with 100,000 Normal Deviates* (The RAND Corporation, The Free Press of Glencoe, Illinois, 1955).

[5]That is, excess disregarding algebraic sign. The absolute value of −7 is 7. Actually some of the listed "absolute excesses" are really deficits; but we are interested here only in the amount by which heads and tails fail to equal each other.

draws closer and closer to the value 0.500. The heads and tails are "balancing out."

But now look at the last column. The absolute excess of heads over tails starts with the value 8, momentarily dips to 2, and then increases to 32 and then to 42, as more and more tosses are made. If this table were extended to 100,000 tosses, or 1,000,000 tosses, the *ratio* could confidently be expected to be still nearer to 1/2, but the absolute excess might grow a good deal. If, as the number of tosses keeps growing, the absolute excess of the appearance of one face also grows, then clearly longer and longer strings of uninterrupted occurrences of one face or the other can be accommodated in the series of results without interfering with the fact that, in the ratio sense, the experiment is becoming better and better behaved.

We will see presently that it is a characteristic feature of any such series of probability trials that it tends to become better and better behaved in the *ratio* sense, but worse and worse behaved, wilder and wilder, in the *absolute* sense.

We also will have, in another chapter, some comments about uninterrupted "runs" in probability experiments; and these remarks should assist you further in grasping and accepting the correct view concerning this basic point: that, in a series of independent trials with fixed probability, the outcome in a given trial does not in the least depend on what the outcome has been in the preceding trials.

The theory of probability furnishes the quantitative and analytical tool for studying the nature of the smoothing-out process which so often occurs when an experiment or trial is repeated a large number of times. In the next two chapters we will consider together some of the principal results that probability theory furnishes for understanding this tendency for irregularities to smooth out in the long run.

Chapter IX

VARIABILITY AND CHEBYCHEV'S THEOREM

It doesn't depend on size, or a cow would catch a rabbit.

PENNSYLVANIA GERMAN PROVERB

Variability

When performing a probability experiment, one does not, of course, expect exactly the same thing to happen on every trial, exactly the same numerical value to result from each measurement, etc. If the result of each trial is a number (such as the number of heads when a coin is tossed 100 times), then with several or many such numbers to consider, one is specially interested in two aspects of this set: First of all, what is the average value of the numbers; and secondly, how do the individual numbers distribute themselves around the average. Do they cluster closely about the average, or do they scatter widely?

The kind of average one usually[1] is interested in is simply the familiar *numerical average* or, as it is usually technically called, the *mean value* obtained by adding all the numbers and dividing by the number

[1]There are other kinds of average, useful in special circumstances.

of the numbers. In this section I want to introduce you to a highly useful way of describing and measuring the amount of *scatter* of the numbers about the mean value.

If a probability experiment has resulted in a sample[2] of n values $x_1, x_2, x_3 \ldots x_n$, then the mean of these values, usually denoted[3] by $\bar{x}$ (called "x-bar"), is simply

$$\bar{x} = \frac{x_1 + x_2 + x_3 + \cdots + x_n}{n} \tag{15}$$

whereas the so-called *variance* of the sample — this variance being the quantity that has proved most useful in characterizing the variability or spread of the numbers $x_1, x_2 \ldots x_n$ about their mean — is defined to be

$$s_x^2 = \frac{(x_1 - \bar{x})^2 + (x_2 - \bar{x})^2 + \cdots + (x_n - \bar{x})^2}{n} \tag{16}$$

You doubtless are curious why, in describing variability or scatter, it was agreed to deal not with the amounts $(x_1 - \bar{x})$, $(x_2 - \bar{x})$, etc. by which the separate values differ from the mean, but rather to deal with the square of these amounts. If you were shooting at a target and missed the bull's-eye, on one shot, by 2″ to the right (which, mathematically, is a *plus* miss), and on the second shot by 2″ on the left (which, mathematically, is a *minus* miss), you would readily agree that it was not fair or sensible to add the $+2''$ and the $-2''$ and obtain zero inches — indicating for the two combined shots no miss at all.

[2] "Sample," because this particular set of n values is only a sample of what we could get if we repeated the experiment for many more trials.

[3] Don't confuse this with the notation E for "*not-E*"!

You might think the easiest thing would be to disregard the algebraic sign and call both simply misses by 2″. But "disregarding the sign" is accomplished in mathematics by "taking the absolute value." The absolute value is denoted by vertical bars, thus

$$\text{absolute value of } +2 = |+2| = 2$$

$$\text{absolute value of } -2 = |-2| = 2$$

But it turns out that absolute value is an awkward thing to work with. For example

$$|7| + |-3| = 10$$

whereas

$$(7) + (-3) = 7 - 3 = 4$$

This is one of the reasons why, when Lady Luck aims at a result but misses, it is handier to deal with the *squares* of her misses, rather than with the first power. The square of a negative number is, of course, a positive number; so squaring gets rid of the difficulty of positive and negative misses, improperly canceling each other out. There are other important reasons why the *square* of the misses leads to a useful measure of scatter; but we have to wait for them to appear later.

The positive square root of this sample variance s_x^2 is, of course,

$$s_x = +\sqrt{\frac{(x_1 - \bar{x})^2 + (x_2 - \bar{x})^2 + \cdots + (x_n - \bar{x})^2}{n}} \tag{17}$$

and it is called the *standard deviation* of the sample.

If there is a theoretical model for the experiment in question, and if we call X the variable which in the model plays the same role that x does in the real experiment, then in the model one defines μ (Greek letter *mu*) the mean value of X, as the expectation of X, namely

$$\mu = E(X) \tag{18}$$

and then defines the variance σ^2 (Greek letter *sigma*) of the theoretical random variable as the expected value of the squares of the deviations of X from μ, thus,

$$\sigma_x^2 = E[(X - \mu)^2] \tag{19}$$

We saw, in Chapter VII, that the expectation in a quantity tends to be more and more closely equal, as the trials increase, to the average or mean value of the differing results. Thus the quantities (18) and (19) are related to the theoretical variable X of the model just as are the quantities (15) and (16) to the actual experimental values $x_1, x_2 \ldots x_n$.

Thus σ_X, the positive square root of the variance, has the value

$$\sigma_X = +\sqrt{E[(X - \mu)^2]} \tag{20}$$

and is called the *standard deviation.*

Although s_x seems to be defined in a rather fancy way, do not lose sight of the fact that s_x is just a kind of average of the misses away from the target value. It is often called a "root mean square" average, and for obvious reasons.

If there are only three values in your sample — $x_1 = 3$, $x_2 = 5$, and $x_3 = 7$ (a rather ridiculously small sample) — then $\bar{x}$ would be

$$\bar{x} = \frac{3 + 5 + 7}{3} = 5$$

and σ_X would have the value,

$$\sigma_X = \sqrt{\frac{(3-5)^2 + (5-5)^2 + (7-5)^2}{3}}$$

$$= \sqrt{\frac{4+0+4}{3}} = \sqrt{2.67} = 1.63$$

This, you must agree, is a reasonable sort of over-all measure of the combined miss of the three shots, since one "shot," namely 5, is right on the average $\bar{x} = 5$, whereas the other two "shots" both miss by 2, and an over-all measure of the miss clearly should be between 0 and 2, and nearer to 2 since there were two misses of 2.

If the model is a good one, and if we have a good-size and fair sample of realized values $x_1, x_2, \ldots x_n$, then we can expect the sample variance as computed from the experimental values in (16) to be near in value to the theoretical value (19) as calculated from the model. In texts on statistical theory you would find a good deal of discussion about the relation between these two values of variance — one the actual variance of a realized sample and the other the theoretical variance of all the values that *would* be experienced if one could carry out an indefinitely large number of trials of the model.

Before going on, perhaps we should look at a very simple numerical example, to get more of a feel for the *mean*, the *variance*, and the *standard deviation*.

Suppose a coin is tossed 10 times and the number x_1 of heads recorded: and suppose this experiment is repeated 8 times, to produce numbers $x_2, x_3, \ldots x_8$. Suppose these right numbers are, in fact[4]

9, 4, 7, 5, 6, 4, 5, 6

[4]I did the experiment, and these are the actual results. You ought to try it, too.

The mean of these eight numbers, their sum being 46, is

$$\bar{x} = \frac{46}{8} = 5.75$$

The variance is

$$s_x^2 = \frac{(9 - 5.75)^2 + (4 - 5.75)^2 + \cdots + (6 - 5.75)^2}{8}$$

$$= \frac{19.50}{8} = 2.44$$

and thus the standard deviation is

$$s_x = \sqrt{2.44} = 1.56$$

This is larger than the value 1.25, which is the average of the deviation from the mean regardless of sign. This is, of course, inevitable because the standard deviation, which involves the *squares* of the deviations, is specially influenced by the larger deviations. In this little experiment the somewhat aberrant result of 9 heads on the first ten tosses (this really happened, please remember) gives a deviation of 3.25 and thus the square, 10.56, contributes more to the standard deviation than do all the other nine results put together.

Chebychev's Theorem

Having defined the *variance* and *standard deviation* of any set of numbers produced by a probability experiment or associated with a probability model, we now can consider a very remarkable theorem obtained, long ago, by the Russian mathematician Chebychev.[5]

[5]Pafnutiy Lvovich Chebychev (1821–1894) was a professor at the University of St. Petersburg and one of Russia's greatest mathematicians.

FIG. 19

We are going to talk about a model. Each trial with this model produces a specific value x_i of the theoretical variable X (the model might be a die, for example; X would designate the number that might appear when the die is rolled, and $x_5 = 3$ might be the face that actually appears on the fifth roll).

We will be dealing with a model of such a sort that X can take on only a definite and finite number of values $x_1, x_2, \ldots x_n$. And we will write, as always, the probability that X does in fact turn out to be x_1 as $P(x_1)$. Similarly $P(x_2), P(x_3), \ldots P(x_n)$ are the probabilities of occurrence of the other possible values $x_2, x_3, \ldots x_n$ of X.

First, we want the expected value of X. That is (see equation 18, page 190),

$$\mu = E(X) = P(x_1) \cdot x_1 + P(x_2) \cdot x_2 + \cdots + P(x_n) \cdot x_n$$

Notice that

$$P(x_1) + P(x_2) + \cdots + P(x_n) = 1$$

because $x_1, x_2 \ldots x_n$ are all the possible values X can assume, and therefore one of them *must* occur (probability of certainty = 1).

Suppose we now rearrange the list of x values in order of their increasing size. Suppose, just for example, that x_7 is the smallest one, x_3 is the next smallest . . . x_2 the next largest, and x_9 the largest. Then in order of increasing size we would have[6]

$$\underbrace{x_7, x_3, \ldots x_i, \qquad\qquad x_{i+1} \ldots x_2, x_9.}$$

The expected value μ is between some entry x_i and the next largest one, x_{i+1}

The expected value μ is a number which in general is not equal to any one of these values; but I have indicated that the mean μ would have a value between two successive entries in this ordered line, usually somewhere near the middle of the line.

This probability experiment (and note the power and generality of what we are doing; for we haven't in the least restricted the nature of the experiment) is, so to speak, "aiming" at the expected value μ which we know will, for a lengthy experiment, be pretty close to the *mean* value. The various numbers x_1, x_2, etc., which can be realized miss the aiming points by various amounts. The realized numbers which are close to μ we would reasonably expect to have higher probabilities than the numbers far away from μ. If the experiment is "aiming" at μ, it ought to have a better chance to come near μ than to miss by a lot.

It is natural, therefore, to be curious as to how heavily, near μ, the probability is concentrated.

[6]The subscript "i" is frequently used, in a situation like this, to denote *some* one of the x's, without specifying (it can't be done in general) *which* one.

Remember that we have a total amount of probability equal to 1, which is distributed among all the realizable values $x_1 \ldots x_n$. Suppose we think of a group of realizable values which are nearer to μ than some certain distance that we will specify. How much of our available probability is concentrated on this particular set of values?

Remembering that we are talking about any experiment you care to think of, with any distribution of the available amount of probabilities over the various possible outcomes, you might very naturally think that the question just asked cannot be answered.

But it can. And the answer will at once convince you that the definition, a few pages back, of the standard deviation, was a real inspiration. For Chebychev proved that at least the amount

$$1 - \frac{1}{h_2} \qquad (21)$$

of the total probability (which of course is equal to 1) is associated with—or we might say allotted to—the realizable values whose distance from the mean is equal to or less than h times the standard deviation σ_X of the theoretical variable X of which $x_1, x_2 \ldots x_n$ are the values realizable in the experiment in question.

Where did h come from in the statement of this theorem? It came right out of the air! The theorem doesn't care a particle what the value of h is. You can choose any h you please.[7]

[7]Any value of h, that is, greater than unity. Values of h smaller than unity lead to negative values of $1 - 1/h^2$, and it doesn't make any sense to talk about a "negative fraction of the measurements."

If you choose $h = 2$, the expression (21) and the theorem say that at least

$$1 - \frac{1}{4} = \frac{3}{4}$$

of the probability is concentrated on realizable values which are located not further than $2\sigma_X$ from the mean value.

If you choose $h = 3$, then you learn that at least 8/9 of the probability is allocated to outcomes not further than $3\sigma_X$ from the mean value μ.

If you choose $h = 5$, then you learn that at least 24/25 of the probability is allocated to outcomes not more than $5\sigma_X$ from the mean value μ.

We have been talking about the amount of probability associated with values clustered near (more specifically, at distances not greater than $h\sigma_X$ from) the mean μ. The theorem has an even simpler sounding form when we speak of the probability associated with values *further from* the mean than $h\sigma_X$. For we can say that the worst that can happen is that the fraction $1/h^2$ of the total probability (namely unity) is associated with realizable values whose distance from the mean μ is more than $h\sigma_X$.

So far in this section we have been talking about the values $x_1, x_2 \ldots x_n$ which, *in a model*, the variable X can assume, with probabilities $P(x_1), P(x_2) \ldots P(x_n)$ that these values will be assumed. You may well be saying, "Wait a minute! If I knew all the values $x_1, x_2 \ldots x_n$ which X can assume, and if I knew all the probabilities $P(x_1)$, etc. — and if I didn't, then I couldn't compute μ and σ_X — then why am I supposed to be impressed by a theorem which tells me how tightly the probability is clustered around the mean? Well, I *know* all the probabilities; so I can, so to speak, just tell by looking at them."

The answer to this lies in the generality of the result. The theorem is proved not by testing special numerical cases,[8] but by means of general reasoning which might occur in some special case. The theorem is a concentrated summary of certain features of *all possible cases.* In a specific case, with the values before you, you can indeed see directly without the theorem — for this one case — how tightly the probability is clustered around the mean. The value of the theorem lies in the fact that it is a general result, true for all cases, whether experienced or not.

Chebychev's theorem as we have just stated it holds for the model; an analogous theorem holds for the actual facts as they are produced in a probability experiment. Suppose you have an actual sample of values $x_1, x_2, x_3 \ldots x_n$ which have resulted from n trials of a probability experiment. We may not know (and for the statements I am about to make do not need to know) *anything whatsoever about the underlying probabilities.* Nevertheless, we can readily compute

$$\bar{x} = \frac{x_1 + x_2 + \cdots + x_n}{n}$$

and

$$s_x = +\sqrt{\frac{(x_1 - \bar{x})^2 + (x_2 - \bar{x})^2 + \cdots + (x_n - \bar{x})^2}{n}}$$

[8]In general, you can't prove theorems by testing special cases. You can, however, often disprove certain general statements by producing a single illustration to the contrary.

and Chebychev's theorem assures us that at least the fraction

$$1 - \frac{1}{h^2}$$

of the measurements differs from the mean by amounts which are equal to or smaller than hs_x.

Suppose I toss an object which can land in only two positions, one of which I will call "up" and the other of which I will call "down." You may wish to think of this as being a round solid cone (see Fig. 20) which can land "upright," resting on its base, or "down," lying on its side.

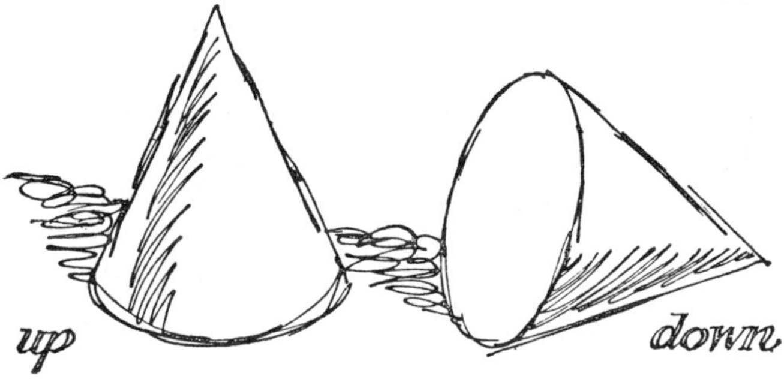

FIG. 20

If the altitude of the cone is small compared to the diameter of the base, then it would be pretty likely to land "up" on its base: but if the base is small and the altitude large, then the cone would be very likely to land "down" on its side. The probabilities for "up" could clearly be given any desired value between 0 and 1 by adjusting the design. Suppose we don't know the design, and hence have no knowledge of the probabilities operating.

Someone "tosses" this object 10 times, giving it a good twist. It lands "up" 3 of the 10 times. In the next set of 10 tosses it comes up 5. In the third set it comes up 2, in the fourth set 4, and in the fifth set 4 again.

We thus have 5 trials of a probability experiment which have resulted in values $x_1 = 3$, $x_2 = 5$, $x_3 = 2$, $x_4 = 4$, $x_5 = 4$. Then

$$\bar{x} = \frac{3 + 5 + 2 + 4 + 4}{5} = \frac{18}{5} = 3.6$$

$$s_x = +\sqrt{\frac{(3 - 3.6)^2 + (5 - 3.6)^2 + (2 - 3.6)^2 + (4 - 3.6)^2 + (4 - 3.6)^2}{5}}$$

$$= +\sqrt{\frac{0.36 + 1.96 + 2.56 + 0.16 + 0.16}{5}}$$

$$= +\sqrt{\frac{5.20}{5}} = 1.02$$

Suppose now we choose $h = 2$. The theorem says that

$$1 - \frac{1}{4} = \frac{3}{4} \text{ or 75 per cent}$$

of the values should have a distance equal to or less than $hs_x = 2(1.02) = 2.04$ away from $\mu = 3.6$. A band 2.04 either side of 3.6 extends from 1.56 to 5.64. If we display the realized values down in order of increasing size we have Fig. 21.

The theorem is satisfied, for in fact *all* the values are located in the interval set up by the theorem.

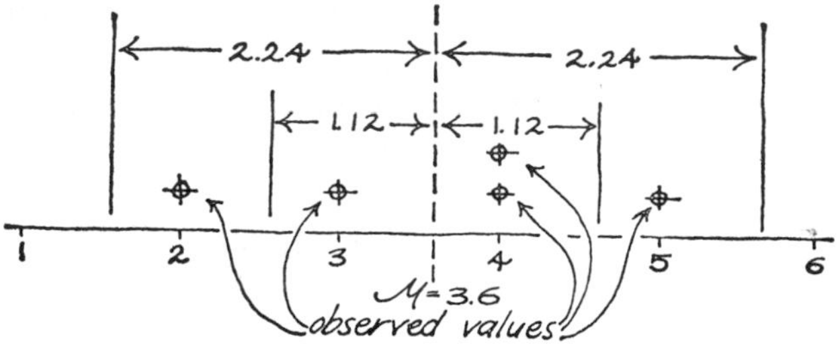

FIG. 21

Suppose we chose $h = 1.1$ so that the theorem says that at least

$$1 - \frac{1}{(1.1)^2} = 1 - \frac{1}{1.21} = 0.17 \text{ or } 17 \text{ per cent}$$

of the values lie not further from $\mu = 3.6$ than $hs_x = (1.1)(1.02) = 1.12$. This sets up the band from $3.6 - 1.12 = 2.48$ to $3.6 + 1.12 = 4.72$. Within this band are, in fact, 3 of the 5 values or 60 per cent of them, thus well satisfying the statement that there will be at least 18 per cent.

Chebychev's theorem gives overly conservative safe lower limits for the percentage of the measurements which lie within various bands, with half-width measured in multiples of the standard variation, about the mean. For the band from $\bar{x} - s$ to $\bar{x} + s$ Chebychev's theorem gives only the trivial and useless result that at least 0 per cent of the values will be in this band. We will see later that this statement can be vastly improved; and that in a good many rather usual circumstances about 68 per cent of the values will be within a standard deviation from the mean.

Inside the band $\bar{x} - 2s$ to $\bar{x} + 2s$ Chebychev assures us that there will fall at least 75 per cent of the values; whereas the usual case, to which we will come later, is that about 95 per cent lie in this band.

Inside[9] the band $\bar{x} - 3s$ to $\bar{x} + 3s$ Chebychev assures us that there will fall at least 89 per cent of the values; whereas under usual probability conditions about 99.7 per cent, or nearly all, of the values lie within this triple-standard deviation band.

From these examples you can see that the moment anyone tries to illustrate Chebychev's Theorem by giving a numerical case, he at that moment condemns the theorem to triviality — almost empty triviality.

For if you have enough concrete information to enable you to use Chebychev's Theorem to calculate the actual numerical value of the probability spread over a certain interval, then you already know so much that there is no use to employ the Theorem. You can then just add up the numerical probabilities and be in a position to make a statement more precise than the estimate you could get from the Theorem.

So the Theorem really isn't much good, you conclude? So it merely is a professional relic that the mathematicians insist on mentioning for sentimental reasons?

Not at all. It is perfectly true that Chebychev's Theorem is not useful for making practical numerical estimates. But it is exceedingly useful (as we will see in Chapter XI) for theoretical purposes — for drawing conclusions in very broad and general circumstances.

This is not a particularly unusual situation. In mathematics a theorem which is exceedingly *general*,

[9]Strictly, "inside or at the extreme limits of."

which holds even if one is lacking a great deal of detailed information, is a most powerful and useful tool, even though it often cannot, in the nature of things, be a very *sharp* tool. If you want results numerically correct to six decimal places, then you must have the values — often to a still greater precision — of all the data that enter in. Then you have to do a lot of rather tedious and dull work, and you come out with an accurate result. But in contrast, it is often much more useful — and almost always a lot more fun — to have methods and results of very great generality, applicable with very few restrictions to a wide variety of circumstances, even though the results are not very precise.

I doubtless could, under a dissecting microscope with suitable precision controls, cut one special and suitable blade of grass to a length of 0.57 inches, correct to ±0.0001 inches. In contrast the rotary blade on my Gravely tractor, much more powerful

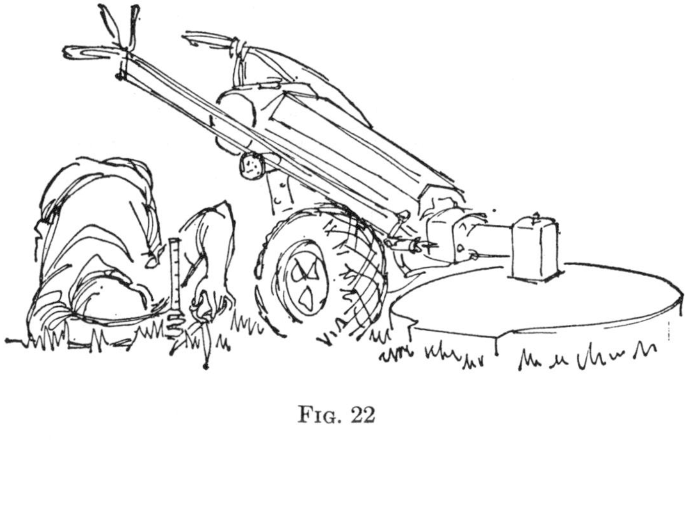

FIG. 22

although less precise, will cut a yard-wide swath to a reasonably uniform height; and it can deal with such a wide variety of grasses, weeds, and even shrubs and small trees, that I hardly have to bother at all about what I am applying it to.

Chapter X

BINOMIAL EXPERIMENTS

. . . but time and chance happeneth to them all.
ECCLESIASTES 9:11

Binomial Experiments

A great deal of useful probability theory results from studying a series of trials of an experiment confined to two outcomes which we may conveniently refer to (without implying that we really want the one and dislike or fear the other) as *success* or *failure*. The probability of success we will designate as p and the probability of failure as q. Since there are only these two mutually exclusive outcomes,

$$p + q = 1$$
$$q = 1 - p$$

We suppose that the successive trials are completely independent, one of another, and that p, and hence q also, does not change at all from trial to trial.

The experiment might be a long series of coin-tossings, in which case we would have $p = q = 1/2$. Or it might be the rolling of a die in the hopes of obtaining the face 5, say, in which case p would be 1/6 and q would be 5/6. Or, of course, it might be repeated trials of a more serious and important nature.

What is the probability that, in a series of N trials, there will be exactly n *successes* and $m = N - n$ failures? Since the individual trials are independent, we can apply the theorem of compound probability (equation 10, page 111) to write down

$$\underbrace{p \times p \times p \cdots p}_{n \text{ factors}} \times \underbrace{q \times q \times q \cdots q}_{m \text{ factors}}$$

$$= p^n q^m \tag{22}$$

that is the probability of getting m successes in an uninterrupted row, followed by m failures in an uninterrupted row.

Exactly the same quantity expresses the probability of getting n successes and m failures in *any one* specified order. But we do not require that the successes and failures come in any special order. We merely require that there be, in all, n of the former and m of the latter.

All the different orders of occurrence of successes and failures, moreover, are mutually exclusive events. If you *do* get one order, you *do not* get any other. Thus by the theorem of total probability (equation 9, page 110) we will have the probability of getting one or another of all the orders if we add expression (22) to itself until the number of terms added is equal to the number of possible orders of N things n of which are alike and of one kind (the n successes) and m of which are alike and of a second kind (the m failures).

Back in Chapter IV it was mentioned (and at the end of the chapter proved) that the formula for the number of orders or permutations of N things, n

of which are alike and of one kind, and $m = N - n$ of which are alike and of a second kind, is

$$\frac{N!}{n!\,m!} = \frac{N!}{n!\,(N-n)!}$$

But (see equation 6, p. 94) this same quantity expresses the number of combinations $\binom{N}{n}$ of N things taken n at a time.

That is, the number of *orders* or *permutations* of N things, n of which are alike and of one kind, the remaining $m = N - n$ being alike and of another kind, can be written, if we choose to do so, as the number of *combinations* of N things taken n at a time, or equally well as the number of combinations of N things taken $m = N - n$ at a time.

Indeed, it is not hard to argue directly (rather than in a roundabout way, as I have just done) that the total number of *orders* (which certainly sounds like a problem in permutations) can be computed equally well in terms of the expression for the number of *combinations* of N things taken n at a time.

For suppose you think of the N trials all labeled as Trial #1, Trial #2, etc.: or simply as $T_1, T_2, T_3 \ldots T_n$. I pick out of these N "objects," which we have artificially made into distinct "objects" by labeling them, a group or combination of n, and agree that these n are all to be successes, and the rest, $N - n = m$ in number, necessarily failures. This choice, if I wrote it down, might look like this:

$$\underbrace{T_3,\ T_7,\ T_{10},\ T_{13}\ \ldots}_{\substack{n \text{ trials which are} \\ \text{agreed to be successes}}} \qquad \underbrace{T_1,\ T_2,\ T_4,\ T_5,\ T_6,\ T_8\ \ldots}_{\substack{m \text{ trials—all the rest—which} \\ \text{are agreed to be failures}}}$$

If these were rearranged to put them all in serial order, there would then be written down

$$T_1,\ T_2,\ T_3^*,\ T_4,\ T_5,\ T_6,\ T_7^*,\ T_8 \ldots$$

(I have put an asterisk on the successes), and I can think of the order as one in which n successes and m failures might occur.

If you took another group or *combination* of n of the T's, then you would, on serial rearrangement, have another of the possible orders. Each group so chosen leads to a different order; and by taking all possible groups of n, you end up with all possible orders of the successes and failures.

This little exercise in reasoning is put in only to strengthen your muscles. For we already knew that

> The number of *permutations* of N things, n of which are alike and of one kind, and $m = N - n$ of which are alike and of a second kind
>
> $$= \frac{N!}{n!\,(N-n)!}$$
>
> The number of *combinations* of N things taken n at a time
>
> $$\binom{N}{n} = \frac{N!}{n!\,(N-n)!}$$

and since the two quantities described in words are equal to the same thing, they must of course be equal to each other. But even so, it is good for you to reason out why a problem in *permutations* may be re-expressed as a problem in *combinations*.

We must not lose track of where we are. Back in expression (22) we had the probability for any one specified order of n successes and $m = N - n$ failures. We know that to obtain the total probability of n successes and m failures in any order we must add expression (22) to itself until we have as many terms as there are possible orders, or, more simply, that we must multiply expression (22) by the number of orders. We now know what the number of orders is. So the total probability of n successes and $m = N - n$ failures in N independent trials of an event for which the probability of success is p on each trial, the probability of failure on each trial being $q = 1 - p$, is

$$\binom{N}{n} p^n q^m \tag{23}$$

Whenever the alternative form is useful, we also can write it as

$$\binom{N}{m} p^{N-m} q^m \tag{24}$$

since $\binom{N}{n}$ and $\binom{N}{m}$ are exactly the same thing; namely

$$\frac{N!}{n!\,(N-n)!} = \frac{N!}{m!\,(N-m)!}$$

when, as here, $n + m = N$.

Why "Binomial"?

Perhaps you recall that the title of the present section, a few pages back, was "Binomial Experiments." Why "binomial"?

When you took elementary algebra in high school

(after all, I assume that you had a decent upbringing), you learned that

$$(a + b)^2 = a^2 + 2ab + b^2$$

and that

$$(a + b)^3 = a^3 + 3a^2b + 3ab^2 + b^3$$

To get these very simple formulas you need only to multiply, longhand, as

$$\begin{array}{r} a + b \\ a + b \\ \hline a^2 + ab \quad\quad \\ ab + b^2 \\ \hline a^2 + 2ab + b^2 \end{array}$$

You doubtless also studied — or were exposed to — what the book called the Binomial Theorem. This theorem found its expression in a formula for multiplying $a + b$ by itself not one or two times, but until there were in all N factors.[1] This formula might have been written

$$\begin{aligned}(a + b)^N = a^N &+ \frac{N}{1}a^{N-1}b + \frac{N(N-1)}{1(2)}a^{N-2}b^2 \\ &+ \frac{N(N-1)(N-2)}{1(2)(3)}a^{N-3}b^3 \\ &+ \cdots + b^N \end{aligned} \tag{25}$$

and you discovered for any given value of N, say $N = 5$, that if you kept on writing terms in accordance with the rule given by the first few terms, you eventually ended up with a last term. Thus

$$(a + b)^5 = a^5 + 5a^4b + 10a^3b^2 + 10a^2b^3 + 5ab^4 + b$$

[1]Your high school algebra probably used a small n here; but I have a special reason, which appears in just a few lines, for using the capital letter.

If you had a somewhat more modern high school algebra book, the binomial formula was written

$$(a+b)^N = a^N + \binom{N}{1}a^{N-1}b + \binom{N}{2}a^{N-2}b^2 + \binom{N}{3}a^{N-3}b^3 + \cdots + \binom{N}{N}b^N \qquad (26)$$

where, just as back on page 94, an expression such as

$$\binom{N}{m} = \frac{N!}{m!\,(N-m)!}$$

equals "the number of combinations of N things taken m at a time." The book reminded you that

$$(a+b)^N = \underbrace{(a+b)(a+b)(a+b)\cdots(a+b)}_{N \text{ factors}}$$

and argued that you could get as many product terms involving $a^{N-m}b^m$ as there are ways of selecting b from m of the N factors (and then necessarily selecting a from the remaining $N-m$ factors). The number of ways of selecting m things from N things is

$$\binom{N}{m}$$

so that a perfectly general term[2] in the binomial expansion of $(a+b)^N$ is

$$\binom{N}{m}a^{N-m}b^m \qquad (27)$$

[2]It is not the m'th term as you write out the whole expression but the $m+1$'st terms. Thus notice in $(a+b)^5$ that the term involving b^3 is not the 3rd term, but the 4th.

Let's try out this general formula, to see that it works. What would be the third term in the expansion of $(a + b)$ to the 5th power? To get the *third* term we must set, as the footnote explains, $m = 2$. Then

$$\binom{5}{2} a^{5-2}b^2 = \frac{(5)(4)(3)(2)}{(2)(1)(3)(2)(1)} a^3b^2 = 10a^3b^2$$

and if you will look back just above equation (26), you will see that this is indeed the third term in question.

That is not a very impressive example, but suppose you wanted to know the 99th term of $(a + b)$ raised to the 101st power. It would be a terrible bore to multiply $(a + b)$ by itself 101 times to get the term in question; whereas the general formula gives at once,

$$\binom{101}{98} a^3b^{98} = \frac{(101)(100)(99)}{(3)(2)} a^3b^{98} = 166{,}650\ a^3b^{98}$$

You may be thinking that we have struggled mightily and not gained anything. Is equation (26) any neater or simpler or more useful than equation (25)?

It is, and for several reasons. One of these is that one can neatly write

$$(a + b)^N = \sum_{m=0}^{m=N} \binom{N}{m} a^{N-m}b^m \tag{28}$$

where the big Greek letter *sigma*,

$$\sum_{m=0}^{m=N}$$

with the little notes above and below, means "add up all the values you get, for the expression just following me, by starting out with $m = 0$ and then letting n successively take on all the integral values 1, 2, 3, 4, etc., until finally $m = N$." The letter *sigma* is a very handy one, for it is so easy to remember that *sigma* means *sum up;* and its shape is handy also, with natural places for writing down the little notes that tell you where to start and where to stop.

We have thus seen that when one multiplies out a binomial $(a + b)$ raised to some power, the coefficients of the various powers of a and b — the so-called binomial coefficients — are precisely the numbers

$$\binom{N}{m}$$

Pascal's Arithmetic Triangle

These binomial coefficients are very interesting numbers. Several pages back we noticed that when N is 2, the coefficients are

1 2 1

and when $N = 3$ they are

1 3 3 1

If N is only 1, then $(a + b)^1 = a + b$ and there are only two coefficients, namely,

1 1

and to go back to the very start, $(a + b)^0 = 1$, since any quantity whatsoever raised to the zero power is equal to unity.[3] Thus for the case $N = 0$ there is only a single coefficient, namely

$$1$$

Many years ago Blaise Pascal wrote a lovely little paper called, in English translation, "The Arithmetic Triangle." This was printed in 1654, the year Lady Luck was born, but not published until 1665, after his death. In it he wrote down the binomial coefficients in a sort of triangle with vertex upward, like this

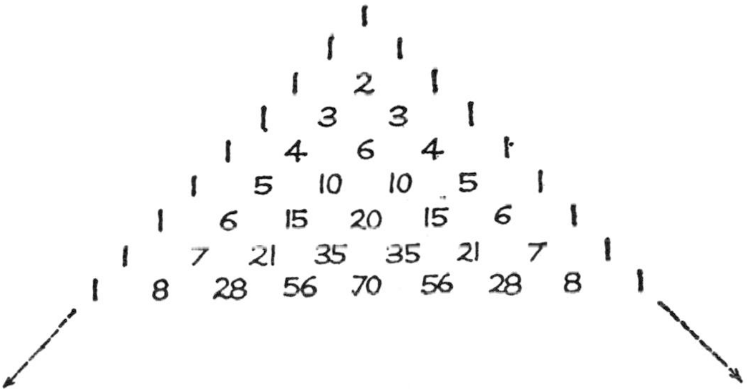

FIG. 23

One obtains any number simply by adding the two numbers just above to the left and to the right. You can tell at once which row you are looking at simply by noting the second number in the row (or the next to the last); and whatever that number is,

[3]This is a convention necessary to adopt if one wants the law of exponents (page 89) to hold in general so that $x^r \cdot x^s = x^{r+s}$ for all values of r and s. For $x^0 \cdot x^r$ should then equal $x^{r+0} = x^r$; hence the multiplier x^0 has to be 1.

the row furnishes you with the binomial coefficients for $(a + b)$ raised to that power. For example, one row starts out with the numbers 1, 7, 21 . . . and the second of these numbers, namely 7, tells you that you are looking at the row which contains the coefficients for raising a binomial to the 7th power. Thus:

$$(a + b)^7 = a^7 + 7a^6b + 21a^5b^2 + 35a^4b^3 + 35a^3b^4 + 21a^2b^5 + 7ab^6 + b^7$$

There are many, many fascinating properties of these numbers known as the binomial coefficients. There is a very large literature on the subject. My friend Fred Mosteller (the senior author of the book I keep referring to) says that Mosteller's Law is: "If you discover a new formula involving the binomial coefficients, it always turns out that it has been previously published."

Having this little excursion into pure algebra behind us, we can now go back to our series of N trials. We know that the probability of n successes and $m = N - n$ failures, namely

$$\binom{N}{m} p^{N-m} q^m$$

[see expression (24)] has exactly the same form as the binomial coefficient given by expression (27). Thus we can state the

Binomial Probability Theorem

If N trials are made each of which results either in "success" or "failure," if the probability of success on each trial is p and of failure on each trial is $q =$

$1 - p$, and if the trials are independent, then the probabilities of all the possible results of such a trial are given by the various terms of the binomial expansion

$$(p+q)^N = p^N + \binom{N}{1}p^{N-1}q + \binom{N}{2}p^{N-2}q^2 + \cdots$$

$$\cdots + \underset{\substack{\uparrow \\ \text{general term}}}{\binom{N}{m}p^{N-m}q^m} + \cdots + \underset{\substack{\uparrow \\ \text{last term}}}{q^N} \qquad (29)$$

and the probability of $n = N - m$ successes and of $m = N - n$ failures is the $m + 1$ term in this expansion. That is

$$\text{Probability} \begin{cases} \text{of } n = N - m \text{ successes} \\ \text{and of } m \text{ failures} \end{cases}$$

$$= \binom{N}{m}p^{N-m}q^m$$

$$= \binom{N}{m}p^n q^m$$

$$= \binom{N}{n}p^n q^m$$

These last three expressions are exactly equal, one to another, by virtue of the fact that $n = N - m$.

If, by all this discussion, we had merely justified the use of the phrase "binomial experiment," then our effort would be rather pointless. But it is in fact true that (29) expresses a very important result, which contributes a great deal to the simplicity of thinking about probabilities in repeated trials, and in carrying out theoretical and practical calculations.

Some Characteristics of Binomial Experiments

First of all, since $p + q = 1$,

$$(p + q)^N = (1)^N = 1 = p^N + \binom{N}{1} p^{N-1}q + \cdots + q^N$$

This is reassuring! For on the right we have the sum of the probabilities of all the mutually exclusive things that can possibly happen when this probability experiment is given N trials. Some one of these *must* happen, so the total probability of all the cases properly comes out to be 1, or certainty.

If $p = q = 1/2$, as it is with coin-tossing, then the binomial expansion simplifies to the form

$$(p + q)^N = \left(\frac{1}{2} + \frac{1}{2}\right)^N = \left(\frac{1}{2}\right)^N \times \left[1 + \binom{N}{1} + \binom{N}{2} + \binom{N}{3} + \cdots + 1\right]$$

and we have (referring to the rows in Pascal's Arithmetic Triangle)

$$\left(\frac{1}{2} + \frac{1}{2}\right)^2 = \frac{1}{4}(1 + 2 + 1)$$

$$\left(\frac{1}{2} + \frac{1}{2}\right)^3 = \frac{1}{8}(1 + 3 + 3 + 1)$$

$$\left(\frac{1}{2} + \frac{1}{2}\right)^4 = \frac{1}{16}(1 + 4 + 6 + 4 + 1)$$

$$\left(\frac{1}{2} + \frac{1}{2}\right)^5 = \frac{1}{32}(1 + 5 + 10 + 10 + 5 + 1)$$

$$\vdots \qquad \vdots$$

The various terms in the parentheses build up in size, reach their maximum at the middle term (if there is a middle one: and otherwise at the *two* central terms), and then decrease.

In other words, when you toss a coin a number of times, the least likely result is to get all heads or all tails, and the probability of other mixtures of heads and tails steadily increases, as we move away from the extreme cases, until it reaches its maximum at an even break between heads and tails (if the number of throws permits an even break), or a shared maximum for the two symmetrical cases nearest to an even break (if the number of tosses is odd).

For values of p and q other than 1/2 the behavior is roughly similar, in that the terms first build up, reach a largest value (or pair of equally large ones) and then decrease.

The largest term in the binomial expansion gives the probability of the outcome of highest probability — that is, the most probable outcome. What is this most probable outcome?

For $p = q = 1/2$ the most probable outcome is at the middle of the expansion; that is, the most probable number of successes in N trials, if the probability for success on a single trial is 1/2, is that where half of the N trials succeed and half fail.[4]

When p and q are not equal, it takes just a little algebra to locate the largest term. The most probable number of successes is r, where

$$\binom{N}{r} p^r q^{N-}$$

[4]This statement is accurate if N is an even number. You can patch it up if N is uneven.

is equal to or larger than its predecessor term (which you would get by replacing r by $r - 1$) and also equal to or larger than its successor term (replace r by $r + 1$). This condition, when worked out, says that the integer r lies between

$$pN + p - 1 \qquad \text{and} \qquad pN + p$$

If you play around with a few numerical examples, you will see that this procedure leads you to a number of successes, as the most probable number of successes, very close to pN. Indeed if pN is an integer, you always get pN as the most probable number. The only reason the algebra seems to be a little messy is that the most probable number of successes must, of course, be an integer, whereas the quantities pN, $pN + p$, etc., are in general not integers.

The result we just obtained is, of course, a very appealing one. If you have a chance of success of 1/8 on each trial, then it seems entirely natural that the most likely number of successes should be one-eighth (or close to one-eighth) of the number of trials.

How probable is the most probable outcome? Let's deal with the simple case in which we toss an even number of times, so that there is one central maximum term in the binomial expansion, corresponding to an even break between heads and tails. It will be simpler if we explicitly recognize that we are talking about an even number N, and so we set $N = 2r$. Then N is sure to be even, although r can be an integer, even or odd.

The middle term in the expansion of $(p + q)^{2r}$ is

$$\binom{2r}{r} p^r q^r$$

or, since we are talking about tossing a coin with $p = q = 1/2$, this is equal to

$$\binom{2r}{r}\left(\frac{1}{2}\right)^{2r}$$

The second factor here gets rapidly smaller and smaller, as r increases. Indeed, it obviously goes down by a factor of 1/4 every time r increases by 1 (or down by a factor of 1/2 every time the number of tosses N increases by 1). Although the binomial coefficient $\binom{2r}{r}$ *increases* as r increases, it does not increase enough to cancel out the decrease in the second factor. In other words, the probability of an exactly even break between heads and tails *decreases* as the number of tosses increases.

It had better! For every time we toss once more we add two more possible cases. We only have a total supply of probability equal to 1 to spread over all the possible cases (corresponding to the fact that one of them must happen); so if we have to feed two more mouths and treat everyone fairly, each gets less than it did before.

If you take the ratio of the even-break probability for two successive values of r (say r and $r + 1$), and divide the second by the first, you get[5]

$$\frac{2r + 1}{2r + 2}$$

This ratio is less than one for any value of r, as we know it must be since we just observed that the even-break probability decreases as the number of tosses increases. But when r is fairly large, this ratio

[5]Do you? It's easy. Work it out.

is only a very little bit less than one. This is also sensible. For if we have a lot of terms (i.e., r is fairly large), then when we increase by two more cases we have increased the total number of cases *relatively* by a very small per cent, and hence decreasing the probability by a small per cent is enough to keep the books balanced at the proper over-all total of "probability of all cases = 1 = certainty."

You may be interested to look at a little table:

TABLE XI

EVEN-BREAK PROBABILITIES WHEN TOSSING COIN $2r$ TIMES

r	*Middle Binomial Coefficient*	2^{2r}	*Probability of r heads and r tails*	*Factor by which successive Even-break Probabilities Decrease*
1	2	4	.500	.75
2	6	16	.375	.83
3	20	64	.313	.87
4	70	256	.272	.90
5	252	1024	.245	.92
6	924	4096	.226	.93
7	3432	16,384	.209	.94
8	12,870	65,536	.196	

The even-break probability for 14 tosses ($r = 7$) is thus .209, whereas the even-break probability for 16 tosses is .196. The latter is .94 as large as the former, which you can easily check is indeed the value of $(2r + 1)/(2r + 2)$.

This table shows that by the time the number of tosses has increased to 16, the even-break probability is decreasing by a factor (0.94) only a little less than unity. On the other hand, the probability of the "extreme" mixtures, all heads or all tails, being

$$\frac{1}{2^{2r}} = \frac{1}{4^r}$$

goes down by a factor of 1/4 every time r increases by 1 (the number of tosses, by 2), no matter how large r is.

Thus comparing the two cases $r = 50$ and $r = 51$, the probability of all heads is only 1/4 as large in the second case as in the first case; whereas the probability of an even break is, in the second case, only reduced by the factor

$$\frac{2r+1}{2r+2} = \frac{101}{102}$$

as compared with the first case.

Since the middle terms do not decrease very rapidly, whereas the first and last terms decrease much more rapidly, you might infer that as the number of tosses increases the probability gets more and more concentrated around the central, "even-break" region. That is not true. Indeed, the opposite is the case.

For $r = 1$ (2 tosses), the diagram which shows the probabilities of all heads, one head and one tail, and all tails looks like Fig. 24.

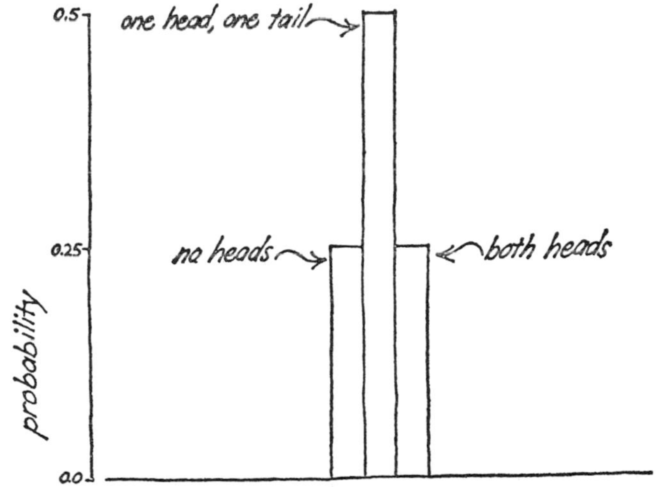

FIG. 24 *Probability of no heads, of one head and one tail, and of two heads when tossing two coins*

For $r = 3$ (6 tosses), it looks like Fig. 25.

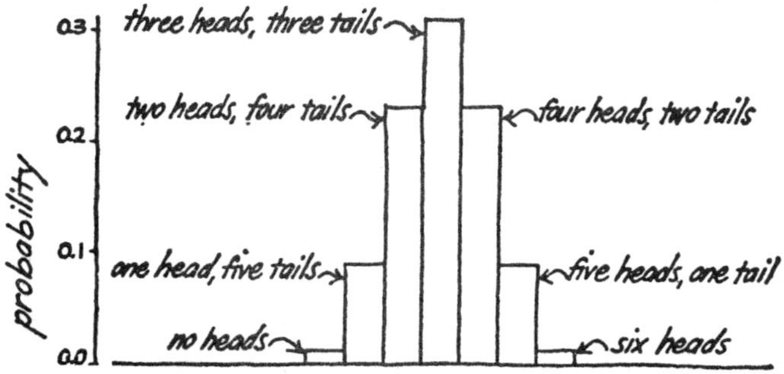

FIG. 25 *Probabilities of 0, 1, 2, 3, 4, 5, or 6 heads when tossing 6 coins*

And for $r = 8$ (16 tosses), it looks like Fig. 26.

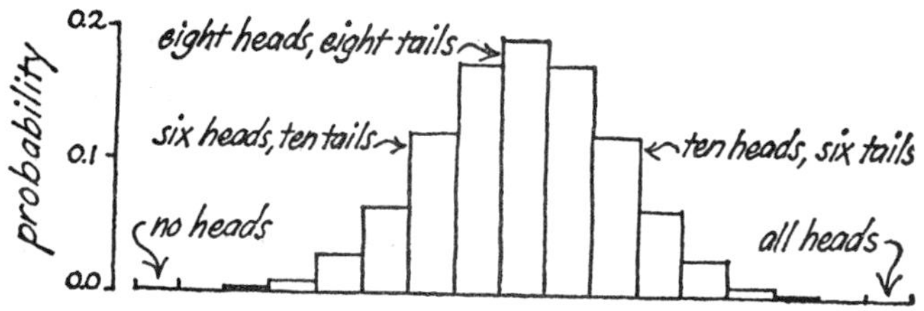

FIG. 26 *Probabilities of 0, 1, 2, 3, 4 · · · 15, 16 heads when tossing 16 coins*

In other words, the picture keeps spreading out (which it must do since the number of tosses is increasing), and it also is *getting flatter*, with less distinction between the probabilities of distributions near the even-break point.

The last table shows that for 16 tosses the probability of an even break is 0.196, so that there is almost one chance in five of getting 8 heads and 8 tails. By the time the number of tosses is increased to 100, the factor $(2r + 1)/(2r + 2)$ keeps whittling down a little at a time the probability of an even break, until the probability of 50 heads and 50 tails is very close indeed to the value 0.08. By 1000 tosses, the probability of an even break[6] is about 0.025; and for 10,000 tosses it is only about 0.008.

What is the *expected* number of successes in a binomial trial? Back on page 172 it was noted that the expectation in the sum of several quantities is the sum of the expectations. When an experiment of N trials is made, the result can be thought of as the sum of the results on each individual trial. On one trial there are but two possibilities. The number of successes is 1 (for which the probability is p), or it is 0 (for which the probability is $q = 1 - p$). Thus the expectation of successes on one trial is

$$1 \cdot p + 0(1 - p) = p$$

and therefore, by addition, on N trials

$$E \text{ (number of successes)} = E(m) = Np \quad (30)$$

Again this is very sensible, and to be expected. The "expected" number of successes (using the word with its precise technical meaning) is exactly the number that a reasonable person would *expect*—namely, that fraction of the trials indicated by the probability (itself a fraction). If the probability is, for example, 1/13, you would *expect* to succeed on one-thirteenth of the total trials.

[6]For a reasonably large number N of tosses the probability of an even break between heads and tails (or as near as N permits you to come to an even break, if N is odd) is about $\dfrac{0.8}{\sqrt{N}}$.

What is the standard deviation in the case of a binomial experiment? We recall, from equation (20) on page 190, that the standard deviation σ is the square root of the expected value of the square of deviations from the expected value.

In our case of a binomial experiment the "value" we are concerned with is m, the number of successes. We have just seen that the expected number of successes is Np. Hence we require

$$\sigma = \sqrt{E(m - Np)^2}$$

Since the algebra involved is a little complicated and tedious — and not particularly interesting — I will just write down the answer. Namely

$$\sigma = \sqrt{Npq} \tag{31}$$

This value will be most useful to us in Chapter XII.

Chapter XI

THE LAW OF LARGE NUMBERS

There is safety in numbers.
AUTHOR UNKNOWN

Bernoulli's Theorem

Several times in the preceding chapters I have made statements along this line: "Since the probability of the event E is $1/3$, one would reasonably expect that E would occur in about one-third of a long run of trials." The purpose of the present chapter is to indicate the basis for statements of the sort.

The early writers on probability theory were unanimous, so far as I know, in defining the probability of an event E in terms of "equally probable outcomes." But it is entirely clear that they also were thinking of the proportion of the cases, in a long run of trials, in which E would actually occur. The first real understanding of the relationship between probability (defined as in equation (1) on page 74), and long-run frequencies of occurrence of events was furnished by James Bernoulli in his treatise *Ars Conjectandi*, or the *Art of Conjecturing*, published posthumously in 1713. It was there that the "law of large numbers" was derived.

The result Bernoulli obtained was the forerunner of a series of more and more powerful, more and more general, laws about probability experiments. These laws are of such central importance to the whole theory that they are ordinarily designated as Central Limit Theorems — the word "Limit" carrying with it the implication of what happens when you make more and more trials, or, as mathematicians say, when you pass to the limit.

I will sketch out a proof of Bernoulli's result, and merely indicate the way in which that pioneering theorem has been improved, for the proofs of the modern substitutes are very complicated. In William Feller's fine treatise on probability theory[1] he had to postpone the general theorem and its proof to his second volume, even though his first volume contained 461 pages of very solid and often advanced mathematics.

Although it was not available, of course, to Bernoulli (for it came a century and a half later than his work), I will make use of the Chebychev inequality which we met back in Chapter IX. You will recall that this inequality, which is not very valuable for studying specific numerical cases, was described as a powerful tool nevertheless for general theoretical purposes. So let's see how that tool works.

We saw, on page 195, that at least the fraction

$$1 - \frac{1}{h^2}$$

of the total probability associated with a random variable is concentrated on cases which are not more than h standard deviations away from the mean.

[1]See concluding note to Chapter III.

Now let us apply this theorem to a binomial experiment. Near the end of the preceding chapter we calculated the mean or expected number of successes for a binomial experiment. We had

$$E(m) = Np$$

so that, inserting a constant factor which does not affect the derivation, and thus rephrasing in terms of the expected *ratio* of successes,

$$E\left(\frac{m}{N}\right) = p$$

Similarly we had

$$E[(m - Np)^2] = Npq$$

so that, for the ratio of successes, $\frac{m}{n}$, the variance is

$$\sigma^2 = E\left[\left(\frac{m}{N} - p\right)^2\right] = \frac{pq}{N}$$

Now remember that when we apply Chebychev's inequality we have the freedom to choose h to be anything we please, just so long as it is greater than unity. It will suit our present purposes to set

$$h = \frac{\sqrt{N}}{t}$$

so that

$$1 - \frac{1}{h^2} = 1 - \frac{t^2}{N}$$

and from here on we have a right to choose t to be anything we please, just so long as it does not exceed $\sqrt{N}$.

Using the value of h we have chosen, then $h\sigma$ has the value

$$\frac{\sqrt{N}}{t}\cdot\sqrt{\frac{pq}{N}} = \frac{\sqrt{pq}}{t}$$

It's important that we have arranged things so that the square root of N cancels out, and our value of $h\sigma$ does not depend on the number of trials.

In terms of these little notational tricks, we can now restate Chebychev's inequality, as it applies to binomial experiments, as follows:

The probability is greater than

$$1 - \frac{t^2}{N}$$

that the actual realized ratio of success m/N does not differ from the probability p by more than

$$\frac{\sqrt{pq}}{t}$$

This still sounds complicated and curious. But just see what we can squeeze out of this remarkable sentence!

First we choose a value of t (which, remember, is at our disposal) *big enough* to make

$$\frac{\sqrt{pq}}{t}$$

"as small as we wish." Then choose N *large enough*[2] so that t^2/N is also "as small as we wish."

[2]Incidentally, this second step assures that, no matter how large we choose t, we now have an N big enough so that the relation between t and $\sqrt{N}$ is a permissible one.

Where are we now? By making the two successive choices of t and N as indicated, we have arrived at the point where we can again restate the Chebychev inequality as follows:

> In a binomial experiment, you can always carry out enough trials (i.e., make N large enough) so that the probability will be just as near certainty as you please (i.e., the probability will be greater than $1 - t^2/N$ and t^2/N is as small as you please) that the ratio of successes m/N differs from the probability p by as little as you please (i.e., by $\sqrt{pq/t}$).

Stated just once more, not so accurately but more simply: *By making enough trials you can essentially secure that the ratio of successes to total trials will closely approximate the probability.*

If the probability for success on each trial is 0.436, then you can make enough trials so that it is *extremely* likely that the actual ratio of successes will be *very* close to 0.436.

To make our illustration more definite, let's say that we want there to be less than one chance in a thousand that the ratio of successes to total trials differs from .436 more than .001.

First we have to make t big enough so that

$$\frac{\sqrt{pq}}{t} < 0.001$$

Putting in the values $p = 0.436$ and $q = 0.564$, this says that t must be greater than 495. Let's take $t = 500$. Then we must make N large enough so that

$$\frac{t^2}{N} < 0.001$$

which requires N be larger than $1000 \cdot t^2$ or 250,000,000.

If you carry out two hundred fifty million trials, then there is in fact less than one chance in a thousand that the success ratio differs from the probability 0.436 by as much as 0.001.

That's admittedly a *lot* of trials. But we never said that the Chebychev inequality was precise and sharp — we only said it was general and powerful.

This is what we have been waiting for! This gives a bridge[3] between the *theoretical, mathematical* probability p, and the actual observed facts of experience. This justifies (or at least starts the justification for) the attractive and natural and very useful idea that a mathematical probability is some kind of guide to the results you can expect actually to realize.

Comments About the Classical Law of Large Numbers

What I have just proved is not exactly what Bernoulli proved — for as noted, he did not have the Chebychev inequality to work with. What Bernoulli actually proved was this:

As each one of the trials becomes longer and longer, the probability tends towards zero that the average success ratio for a large number of such trials will differ from the probability by any preassigned number, however small that number is.

The joker in that sentence is that it applies to the success ratio *as averaged over a large number of trials.* It does *not* give an assurance about the success ratio in one given series of binomial experiments, *nor*, if the

[3]Even though, as we will see in a moment, a somewhat narrow and shaky one.

success ratio in that given series has in fact come very close to p, *does it assure* that the success ratio will keep on being that close (or closer) to p as that particular series of experiments is continued.

The result we have established here with the Chebychev inequality is a "stronger" result than was Bernoulli's original "law of large numbers." For once having agreed on t, then our result holds for *any* N larger than the critical value which makes t^2/N "as small as we please." Thus our result[4] is stronger than Bernoulli's because it applies to one specific series of binomial trials, and because it says that if we continue to a certain number of trials, the success then, *and for all subsequent trials*, will differ from p by an amount which will be less than, and *stay less than*, the small quantity we originally decided upon.

It should be recognized that the theorems stated above — both the Bernoulli form and the Chebychev form — are statements *about the probabilities of events not about events*. If you calculate from the theorem that by making N trials you can make the probability less than one in a thousand that something-or-other will happen, it still remains possible that the first time you carry out the N trials, the something-or-other *does* happen, even though you had calculated it to be very unlikely. If the probability of a certain outcome is only one in a thousand, this does mean after all that it is likely to happen on the average about once in a thousand times. And you just might start out with the unlikely event.

There is another exceedingly fundamental and important aspect of the theorem we have here proved.

[4]I call it "ours" meaning of course that it is yours and mine. It isn't original, of course, by a long, long shot.

Namely, be sure you notice carefully that it describes the good behavior of the *success ratio*, m/N.

If you try to trick the Chebychev inequality into proving a similar statement for m itself, for the actual number of successes, you will discover that no matter how you try to select clever values, first for t and then for N, you always get stuck. The square root of N always turns up in the wrong places.

This, of course, is precisely what was mentioned back in Chapter VIII, when it was pointed out that, as a probability experiment proceeds, the *success ratio* tends to behave *better* and *better*, but the *absolute number of successes* tends to *behave more and more wildly*.

Indeed there is a sort of converse Bernoulli theorem which could be stated something like this:

> By making the number N of trials large enough, you can make as near unity (certainty) as you desire the probability that the actual number m of successes will *deviate from* the expected number *np by as much as you please.*

It is easy to see how the proof of this converse theorem goes. It is easy to show that the probability of the most probable (which is also the expected) number of successes in a binomial experiment is[5] $1/\sqrt{2\pi Npq}$. Hence the sum of the probabilities of the $2r + 1$ outcomes which cluster about the most probable outcome, extending for r cases on either side of the most probable, is less than

$$\frac{2r + 1}{\sqrt{2\pi Npq}}$$

[5]This, in fact, is the expression (see page 223) which reduces to $0.8/\sqrt{N}$ when tossing a coin, with $p = q = 1/2$.

for we would be adding $2r + 1$ quantities, all but one of which are smaller than $\sqrt{2\pi Npq}$.

Now choose r just as large as you please. Having fixed on that choice, then make N large enough so that the above expression is as small as you please.

You have thus determined how many trials will insure that the probability is as near unity (certainty) as you please that the number of successes lies *outside* this $2r + 1$ interval (which was itself chosen as large as you please).

Let us illustrate this with a coin-tossing example. How many times must one toss a coin to make the chances 999 to 1000 that the discrepancy between the number of heads and tails will *exceed* 500? Here $p = q = 1/2$, and $r = 250$.

Thus we must make N large enough so that

$$\frac{2(501)}{\sqrt{2\pi N}} = \frac{399.74}{\sqrt{N}}$$

is less than 1/1000 (so that the probability of being *outside* the interval is 999/1000). This requires that

$$\sqrt{N} > 399{,}740$$

or

$$N > 1.598 \times 10^{1}$$

This is, to be sure, a good lot of tosses! But even so, it does show that the absolute discrepancy between the number of heads and tails tends to get larger as the tosses go on and on, not less and less, as many persons seem to think.

This is so completely fundamental an aspect of probability phenomena that, at the risk of boresome repetition, I say again: As experience accumulates, probability phenomena behave better and better in the ratio sense, and worse and worse in the absolute sense.

Improved Central Limit Theorems

The results we have considered together so far in this chapter would often be referred to as "weak laws of large numbers." The first "strong law of large numbers" was originally and partially discussed by the mathematicians Borel and Hansdorff, and first really formulated and proved in 1917 by Cantelli.

The aspects of strength in the 1917 result, and in the similar but still more powerful theorems which were proved by Lindeberg in 1922, Khintchine in 1929, and still later by Feller and others, are so detailed and so sophisticated that it is not feasible to describe them here. With greater and greater generality and with stronger and stronger assurance, these more powerful laws certify that, as the trials go on and on, more and more closely can you expect the probability predictions to be satisfied, more and more confident can you be that, once satisfied to a certain degree, they will never subsequently go wrong.

Note on Large Numbers

Since this chapter discusses the law of large numbers, it may be appropriate to include a few comments on large numbers themselves.

There are more than 10^{19} molecules in a cubic centimeter of gas under normal conditions; and Sir James Jeans pointed out that when anyone draws a breath it is highly probable that it contains some of the molecules of the dying breath of Julius Caesar.

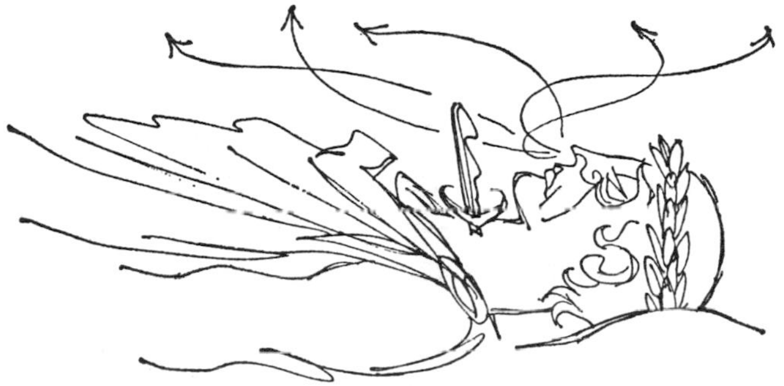

FIG. 27 *Caesar's last breath*

A puzzle known as the Tower of Hanoi involves three pegs, on one of which is a conical pile of round disks, each larger than the one just above it. The game involves moving all the disks to one of the empty pegs, shifting one disk at a time, and never putting a larger disk on a smaller one. The minimum number of moves in which this can be done, with n disks, is $2^n - 1$. In a temple in Benares there is a Tower of Hanoi with 64 disks. The minimum number of moves for this tower is thus

$$2^{64} - 1 = 18,446,744,073,709,551,615 = 10^{19.3}$$

so that if a person made one move a second, night and day, the solution would require about two thousand billion years.

In the lore of India there is a tale about a stone, a cubic mile in size, a million times harder than a diamond. Every million years a holy man visits the stone to give it the lightest possible touch. How long does it take to wear the stone away? On the basis of reasonable estimates of the wear from each touch, this works out to be of the order of 10^{35} years. (I state here the result I have seen printed in several places. But I cannot check the 10^{35}.) If you calculate the number of "atoms" of carbon in a cubic mile of density 10^6 times that of the ordinary diamond, you get a number of the order of 10^{45}. I put the word atom in quotes, because one is really dealing with an interlocked crystal structure: but "the lightest possible touch" would seem to imply removing one atom at each touch. Removing one every 10^6 years then indicates that 10^{51} years would perhaps be required. But don't stick around to see which estimate proves to be correct.)

Archimedes did not have available the neat exponent scheme we use for large numbers, but he devised something equivalent, and estimated the number of grains of sand it would take to fill a sphere with center at the center of the earth, and radius reaching out to the sun. This comes out to be of the order of 10^{51}.

Several years ago, and before the discovery in physics of a whole array of "new" elementary particles, A. S. Eddington gave a strange and powerful argument that, among other things, led to a value for the total number of elementary particles (he was then dealing with electrons and protons) in the universe[6]. This number was $3/2 \cdot 136 \cdot 2^{256}$, or roughly 10^{79}. From one point of view you might well think

[6]"I believe there are 15,747,724,136,275,002,577,605,653, 961,181,555,468,044,717,914,527,116,709,366,231,425,076,185, 631,031,296 protons in the universe and the same number of electrons." Eddington — one of the Tarner Lectures in 1938.

that the total number of particles in the whole universe is as large a number as has any sense. But 10^{79}, as you will see in just a moment, is, relatively speaking, only a rather mildly large number.

At various times writers have spoken of the possible production, by pure chance, of some classic piece of literature by monkeys pecking away at typewriters. There being in Hamlet, for example, something like 27,000 letters and spaces, and there being 35, say, keys on the typewriters (so the chance each time of a monkey's hitting the right key is 1/35), the over-all probability of producing Hamlet by chance is unity divided by 35 raised to the power 27,000, or roughly[7]

$$10^{41,600} = 10^{10^{4.61}}$$

It has not proved feasible to estimate the total number of games of chess; but it has been shown that the number is less than

$$10^{10^{7.5}}$$

There are about nine million books in the Library of Congress, five million in the British Museum, and five million in the National Library of France. Taking into account duplications and books in other languages, it has been estimated that there are of the order of twenty-four million "important" books in the world. If one takes a rough average of 800 pages per book, 500 words per page, and 7 symbols (including spaces and punctuation) per word, this leads to a

[7]Here the exponent of 10 is so large that it is convenient to make use of the fact that $41,600 = 10^{4.61}$ and write the large number with a "two-story" exponent.

guess that about $6 \cdot 10^{13}$ symbols are involved in all these books. To produce *all these* by chance involves hitting the right one of 35 keys on a typewriter $6 \cdot 10^{13}$ times in a row. The probability of this is roughly one divided by[8]

$$10^{100,000,000,000,000} = 10^{10^{14}} = 10^{10^{10^{1.15}}}$$

As contrasted with the probability of *typing* the manuscript of Hamlet by chance, Littlewood has estimated the probability of producing the *written manuscript* of Hamlet by spattering, by pure chance, molecules of ink on paper. This turns out to be of the order of unity divided by

$$10^{10^{20}} = 10^{10^{10^{1.3}}}$$

Thus it is more probable that *all* books could be *typed* by chance than that the *manuscript* of Hamlet could be reproduced by chance.

The velocities of the molecules of gases or solids are distributed in accordance with the Gaussian Law [equation (32)]. Thus in a solid at a certain temperature there is some small chance of finding molecules with velocities much higher than — or much lower than — the mean velocity characteristic of that temperature. Indeed, it is theoretically conceivable that, over a stated period of time, *most* of the molecules might have velocities much lower than the value characteristic of the temperature. This is, roughly, the basis for saying that there is *some* chance that a teakettle of water, put on the stove, will freeze

[8]Since 14 is larger than 10, this leads us for the first time to write a large number in three-story exponent form.

rather than boil. Littlewood has estimated the probability that a celluloid mouse placed in the temperature $2.8 \cdot 10^{12}$ degrees Absolute (which he assumes for hell) would survive without burning for a week. This turns out to be of the order of unity divided by

$$10^{10^{50}} = 10^{10^{10^{1.7}}}$$

The numbers I have mentioned so far are invisible pigmies as compared with some others. All these numbers are, in fact, as nothing compared with a number, called the Skewes number, which is involved in a theorem about the distribution of prime numbers. The Skewes number is

$$10^{10^{10^{10^{1.53}}}}$$

This number, if written out in full, would require roughly

$$10^{10,000,000,000,000,000,000,000,000,000,000,000}$$

digits to write it down.[9]

When a mathematician says "large," he isn't kidding.

* * * * * *

"If six monkeys were set before six typewriters it would be a long time before they produced by mere chance all the written books in the British Museum; but it would not be an infinitely long time." (Extracted from an address given before the British As-

[9]See "Size," by Warren Weaver, *The Atlantic Monthly*, September 1948.

sociation for the Advancement of Science — sometimes referred to, in England, as the British Ass.)

Life is brief, but art is longer
  So the sages say in sooth —
Nothing could be worse or wronger
  Than to doubt this ancient truth.
Endless volumes, larger, fatter
  Prove man's intellectual climb,
But in essence it's a matter
  Just of having lots of time.

Give me half a dozen monkeys,
  Set them to the lettered keys,
And instruct these simian flunkies
  Just to hit them as they please.
Lo! The anthropoid plebeians
  Toiling at their careless plan
Would in course of countless aeons
  Duplicate the lore of man.

Thank you, thank you, men of science!
  Thank you, thank you, British Ass!
I for long have placed reliance
  On the tidbits that you pass.
And this season's nicest chunk is
  Just to sit and think of those
Six imperishable monkeys
  Typing in eternal rows!

Lucio
(in the Manchester *Guardian*)

Chapter XII

DISTRIBUTION FUNCTIONS AND PROBABILITIES

And it is only from Analogy that we conclude the Whole of it to be capable of being reduced into them: only from our seeing, that Part is so.

BISHOP BUTLER, *The Analogy of Religion*

Probability Distributions

Whenever there is an event E which may have outcomes $E_1, E_2, \ldots E_n$ whose probabilities of occurrence are $p_1, p_2, \ldots p_n$, we can speak of the set of probability numbers as the "probability distribution" associated with the various ways in which the event may occur. This is a very natural and sensible terminology, for it refers to the way in which the available supply of probability (namely unity) is *distributed* over the various things that may happen.

A probability distribution may be exhibited in various ways. One obvious way is to show the values in a table. If the event in question is the tossing of two coins and you are interested in what each individual coin is likely to do, the table would look like this:

Outcome	*Probabilty*
E_1 — *HH*	1/4
E_2 — *HT*	1/4
E_3 — *TH*	1/4
E_4 — *TT*	1/4

Or if you are interested only in the number of heads and tails, it would look like this:

Outcome	*Probability*
E_1 — two heads	1/4
E_2 — one head, one tail	1/2
E_3 — two tails	1/4

If the event in question is the tossing of six coins, and if your interest is concentrated on the question "How many heads are there?" then the probability distribution could be exhibited this way:

No. of heads	0	1	2	3	4	5	6
Probability	1/64	6/64	15/64	20/64	15/64	6/64	1/64

These same facts could be depicted graphically. For example, the last table could be replaced by the diagram in Fig. 28.

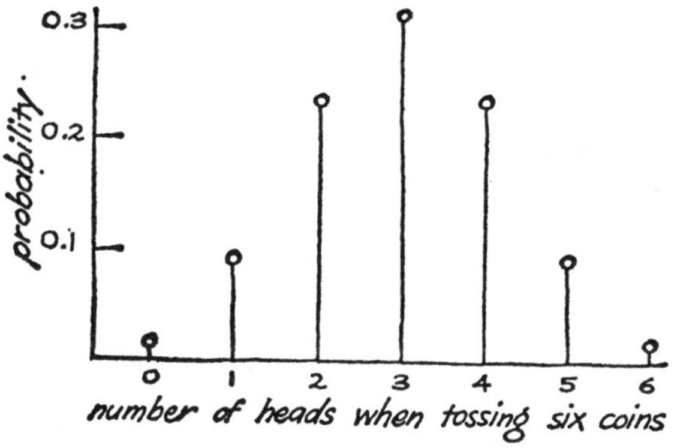

FIG. 28 *Distribution of probabilities as shown by the vertical height of the lines on the graph*

Still another way is to use a diagram such as Fig. 29. Here, just as in the previous diagram, the heights

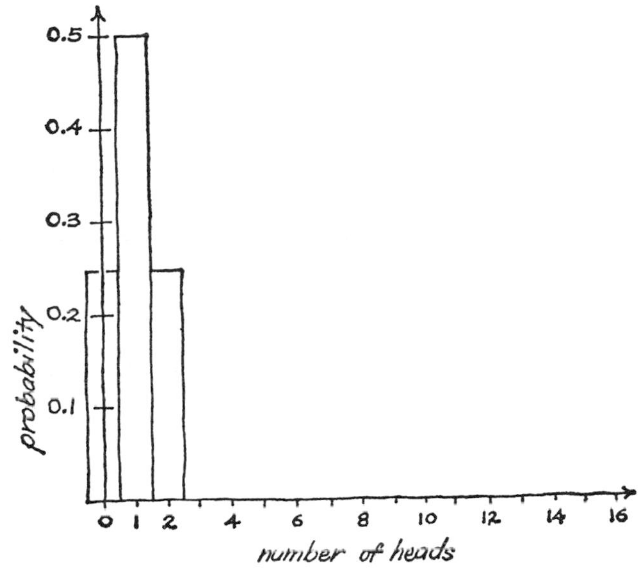

FIG. 29 *Distribution of probabilities measured by the vertical height as well as by the areas of the rectangles (2-coin case)*

over the various integers correspond to the probabilities. But since each of the vertical rectangles is a unit wide at its base, it is true that the *area* of each rectangle (being the unit base times its height) is also a measure of the probability of the number of heads in question. Thus the eye has a double guide — height and area — for judging the relative sizes of the various probabilities depicted.

This idea of a chart in which *area* measures or exhibits probability turns out to be a very useful one. Let's look at another example (Fig. 31), in which not 6 but 16 coins are tossed.

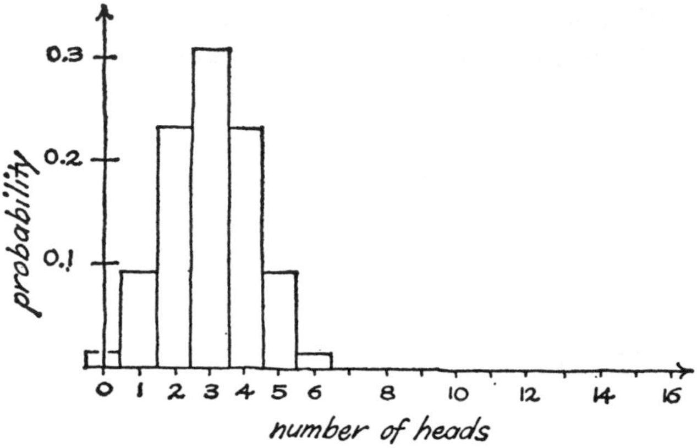

FIG. 30 *Distribution of probabilities measured by the vertical height as well as by the areas of the rectangles (6-coin case)*

Here we have used the same vertical scale as in the preceding chart; and the central largest probability — the break-even probability we talked so much about in Chapter X, which was 20/64 or 0.313 in the previous case of 6 coins — is, of course, smaller, and

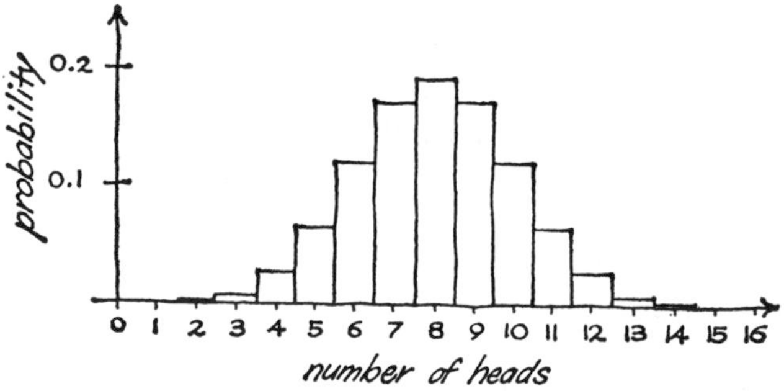

FIG. 31 *Distribution of probabilities measured by the vertical height as well as by the areas of the rectangles (16-coin case)*

in fact 0.196, for the break-even probability of 8 heads and 8 tails in the case of tossing 16 coins.

The chart for 16 coins differs from the chart for 6 in three main respects. First, it is obviously *broader but not so high;* second, it is *flattened out,* for you can see with your eyes that the probability does not fall off so rapidly in the case of 16 coins as you move in either direction from the break-even point; third, its middle has necessarily moved to the right along the x-axis from 3 (the break-even point for 6 coins) to 8 (the break-even point for 16 coins).

On the other hand, the two charts are *exactly alike* in one basic respect: The total area of each (that is, the sum of the areas of all the rectangles) is exactly unity in each case, corresponding of course to the fact that the sum of all the probabilities of the total number of the mutually exclusive cases is equal to 1 or certainty. When you toss a coin any number of times, *some* result *has* to occur!

Normalized Charts

To see what happens to the distribution of probabilities when the number of tosses keeps increasing, it is convenient to modify the successive charts, not in any way altering the information which each contains, but merely changing the vertical scale so that the break-even probability *looks* the same in each instance, and shifting them horizontally so that the break-even point is located similarly for all.

In fact, it is handy to deal not with the actual number of heads but rather with the *discrepancy* or *deviation* in the number of heads away from the break-even number. With this agreement, +2 on the x-axis would mean "two more heads than an even break," so it would correspond to 5 heads in the 6-coin case, but 10 heads in the 16-coin case.

Making these two changes, which ease the comparison between the cases, the 6-coin chart and the 16-coin chart now look like Figs. 32 and 33.

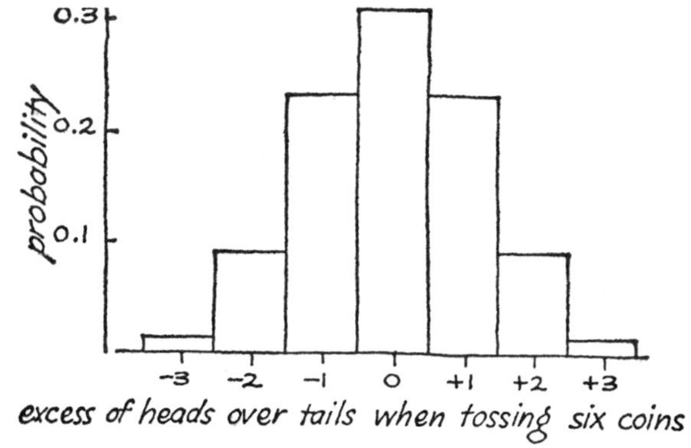

FIG. 32 *Distribution of probabilities for 6 coins expressed in terms of excess of heads*

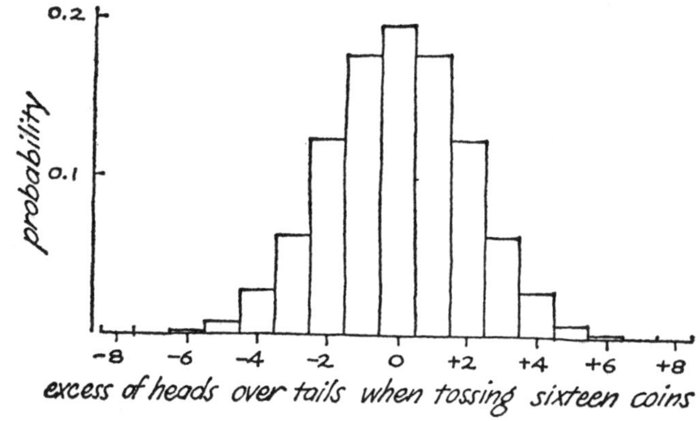

FIG. 33 *Distribution of probabilities for 16 coins expressed in terms of excess of heads*

There are two further changes that, if made in charts of this sort, reveal even more usefully what is really going on as the number of coins, or the number of tosses increases.[1]

The first one of these further changes should not surprise you; for in Chapter IX we have seen that the standard deviation σ is a natural unit for measuring deviations away from the most probable result. In the previous chapter it was noted [equation 31] that the standard deviation for any binomial experiment is $\sqrt{Npq}$. For 6 tosses with $p = q = 1/2$, σ comes out to be 1.22; whereas for 16 tosses, $\sigma = 2.0$.

Now, for one of the two final changes in our chart, we are going to introduce a new horizontal variable which, instead of being the actual discrepancy of heads over tails, will be the discrepancy measured in terms of the standard deviation. This means that the beginning point -3 on the 6-toss chart will become $-3/1.22$, or about -2.45, on the new chart; and the beginning point -8 on the 16-toss chart will become $-8/2 = -4$ on the new chart.

But if we *shrink* the horizontal dimension of our charts by one factor, we ought to *expand* the vertical dimension by the same factor, so that the *area*, which we wish to continue to use as a measure of probability, will not be affected.

Thus the middle height 0.313 for the 6-toss case, and all the other 6-toss heights, are now going to be

[1]Clearly tossing 6 coins once is the same thing as tossing one coin 6 times: so from here on we will speak in terms of the number of tosses of a single coin. It would be awkward to toss 100 coins at once, but perfectly feasible to toss one coin 100 times.

multiplied by 1.22: and similarly all the heights for the 16-toss case are to be multiplied by 2.

This will give us new charts in which, be sure to remember, the vertical scale no longer measures probability. The probabilities are now measured only by the *areas* of the rectangles.

These new charts look like Figs. 34 and 35.

Notice that, when pictured with these "natural" scale factors, based on the standard deviation, the two charts, one for 6 tosses and the other for 16, have now become much more similar. In both cases the rectangles extend out about 3 standard deviations

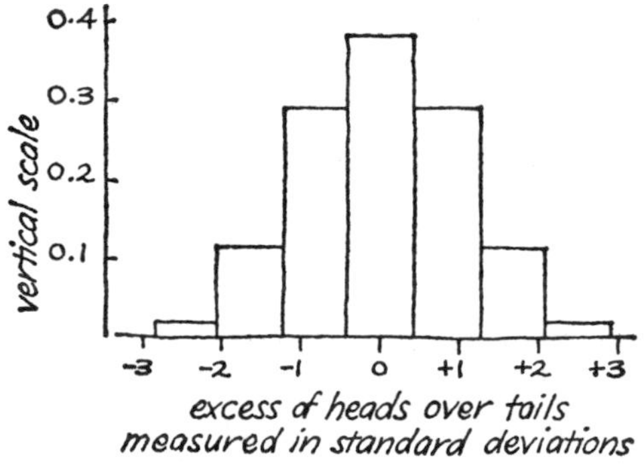

FIG. 34 *Normalized distributions of probabilities, measured now by the areas of the rectangles, not by the vertical scale, for the 6-coin case*

either side of the break-even point. In both cases the maximum height (which, remember, is *not* the probability any more) is about the same (it is 0.383 for the first, and 0.392 for the second). In both cases the diagram slopes away from the central peak at about the same rate.

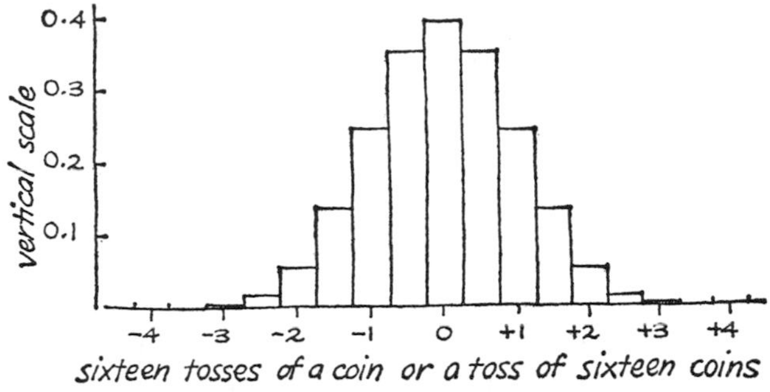

FIG. 35 *Normalized distributions of probabilities, measured now by the areas of the rectangles, not by the vertical scale, for the 16-coin case*

It will therefore not surprise you to look next at an exactly similar chart (Fig. 36), drawn for 100 tosses of a coin.

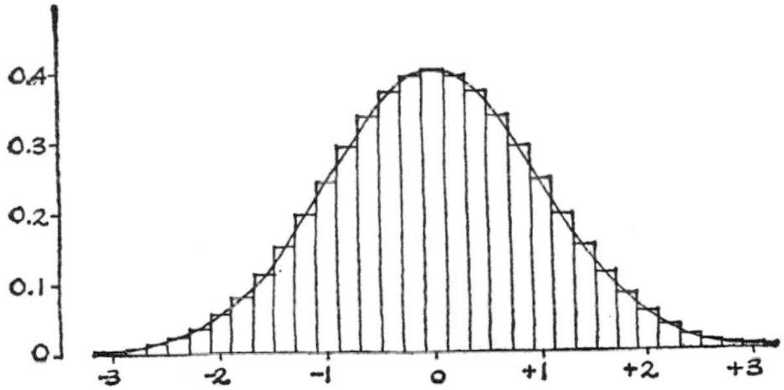

FIG. 36 *Normalized distributions of probabilities, measured now by the areas of the rectangles, not by the vertical scale, for the 100-coin case*

Again you see that the rectangles, whose areas measure the probabilities, extend out to just about 3 standard deviations at either side of the break-even point. Again you see that the maximum height of the central rectangle is about the same as it was in the two preceding charts.[2]

The Normal or Gaussian Distribution

The three charts just shown should make it very reasonable for you to believe that if one tossed a coin more and more and more times, the corresponding charts (when "normalized" as a mathematician would say, by using in each case the standard deviation to shrink the horizontal dimensions and stretch the vertical) would become more and more like a bell-shaped *curve*. And the area under this curve and between any two vertical lines from the x-axis up to the curve, ought to measure the probability of that range of deviation of heads away from the even break, this deviation being measured in terms of the standard deviation as a unit.

This is, of course, precisely correct. It is perfectly possible (although tedious) to draw a chart for 1000 tosses, or even 10,000 tosses. The little horizontal tops of the rectangles would be very short, each being of the length obtained after shrinking a unit length by the standard deviation. Since the standard deviations would be 15.8 and 50.0, for 1000 and 10,000 tosses respectively, the treads on the little stairsteps would be only 0.06 long in the former case,

[2]The maximum height of the central rectangle is 0.400. Since in this case the standard deviation $\sqrt{Npq} = \sqrt{100\cdot\frac{1}{2}\cdot\frac{1}{2}} = \sqrt{25} = 5$, the probability of the even-break of 50 heads and 50 tails is $0.400/5 = 0.080$, giving a good check with value as calculated from the exact binomial expression in Chapter X, page 223.

and only 0.02 long in the second case. Unless you made a rather large drawing, the stairsteps would barely show.

But it would be laboriously foolish to draw these charts for larger and larger numbers of tosses. It's much better to work out a little theory, and replace this jagged diagram with a smooth curve for which you have an equation.

This theory has, of course, been worked out long ago. There are many ways of arriving at the equation for the normal or Gaussian distribution curve. As one way, it really is not very difficult to start out with expression (29) of Chapter X, which gives the probability of exactly m successes in N trials; to write this down for several successive values of m; to add these together; to make use of the approximation for factorials (page 90) in order to get rid of the unpleasant factorials; to make changes in scale and center of symmetry which correspond to those we made above in passing from Figs. 32 and 33 to Figs. 34 and 35; and to let N get bigger and bigger. Then out comes the simple and lovely equation:

$$y = \frac{1}{\sqrt{2\pi}} e^{-x^2/2} \tag{32}$$

This curve is called the normal, or the Gaussian distribution curve, in honor of the great mathematician Karl Friedrich Gauss (1777–1855), who made extensive use of the curve, although it was used earlier in probability theory by de Moivre, and also by Laplace.

Here is the way the normal, or Gaussian, probability curve looks, and you see at once its family relationship with Figures 34, 35, and 36.

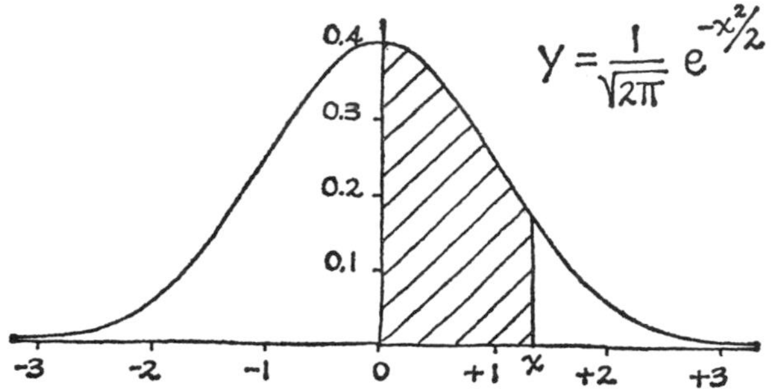

FIG. 37 *The Gaussian or normal distribution curve*

The variable x measures the deviation of the number of successes away from the most probable number of successes pN, and measures this deviation in terms of the standard deviation as a unit. That is,

$$X = \frac{l}{\sigma} = \frac{m - pN}{\sigma}$$

The shaded area is numerically equal to the probability that m will differ from pN by an amount equal to l or less.

The area out to $x = 1$ is 0.3413. Thus the probability that m exceeds pN by less than one times the standard deviation is 0.3413. The probability that m differs (without regard for sign) from pN by less than σ is 2(0.3413) or .6826.

Similarly the probability that m exceeds pN by less than 2σ is 0.4772, this being the area under the curve from $x = 0$ to $x = 2$. The probability that $|m - pN|$ will be less than 2σ is thus .9544.

By the time $x = 3$ the curve is very near the *x-axis*. The area from $x = 0$ to $x = 3$ is 0.4987 so that the probability that $|m - pN|$ will be less than 3σ is 0.9974.

These statements can be summarized by saying, in somewhat figurative language, that 68.2 per cent of the probability is less than one standard deviation away from the most probable result, 95.4 per cent less than 2σ away, and 99.7 per cent less than 3σ away.

Emphasizing the alternative aspect, only in 31.8 per cent of the cases, on the average, will the deviation exceed one standard deviation, only in 4.6 per cent of the cases will it on the average exceed 2σ, and only in 0.3 per cent of the cases will it on the average exceed 3σ.

All these statements, of course, refer to cases in which the normal or Gaussian probability curve applies. The examples to be given presently will clarify this caution.

In carrying out the approximation steps which lead to equation (32) it is possible to make a refinement. Thus by taking the area under the curve from one origin not to the point l/σ, but a slightly more distant point, namely $(l + 1/2)/\sigma$, one gets a slightly more accurate value.

The indicated approximation procedures for arriving at equation (32) work best when p and q are not too far away from 1/2, and also when the maximum deviation l for which the curve is used is restricted to be less than the square root of N. Thus for 100 tosses of a coin, the values 1/2 for both p and q are favorable to the use of the approximation, at least for deviations from the even break which do not exceed $\sqrt{100}$ or 10.

On the other hand, it is hardly to be expected that the symmetrical curve (32) would work very well for the rather unsymmetrical case when one is interested in the number of times one particular face

($p = 1/6$) of a die comes up. And one would hardly expect the curve to work very well for small values of N, or for values of deviations larger than $\sqrt{N}$.

But the curve works better than one has a right to hope. For example, consider our previous case of tossing a coin only 6 times. The probability of getting 2, 3, or 4 heads is, accurately,

$$\frac{15}{64} + \frac{20}{64} + \frac{15}{64} = \frac{50}{64} = 0.781$$

To approximate this same probability by means of the Gaussian curve (which is a brave thing to attempt, for so small a number as $N = 6$), we note that for $N = 6$, $p = q = 1/2$, the standard deviation σ, or $\sqrt{Npq}$, is 1.22: and that, in our problem, l (namely, $4 - 3$) is 1. Hence we require double the area, under the Gaussian curve, from the origin out to 1/1.22, or 0.82.

The area under the Gaussian curve has been tabulated in great detail and accuracy. From the nice little table in the Mosteller book, for example, we read that the area out to 0.82 is equal to .2939 so that the double of this area is, to three decimal places, 0.588. If we make the minor correction referred to, we would take the area out to 1.5/1.22, or 1.23. This leads to the double area value 0.781.

The exact answer to our problem being 0.781, it is really remarkable that the Gaussian curve leads to the fairly good approximation 0.588, and if the correction term is included, to a value which checks the exact value to three decimal places.

The Gaussian curve can equally well be used to estimate the probability that deviations from the most probable number will *exceed* some stated quantity. Thus one can ask in connection with 100 tests of an event for which the probability of success on each trial is 1/5, what is the probability that the realized number of successes will be 24 or *more?* Here $\sigma = \sqrt{100 \cdot 1/5 \cdot 4/5} = 4$, the most probable number of successes is 1/5 times 100 or 20, and we must take the area under that arc of the Gaussian curve that lies to the right of $(24 - 20)/4 = 1.0$ (or if one uses the more precise approximation, to the right of $(24 - 1/2 - 20)/4 = 0.875$.

The area to the right of 0.875 under the Gaussian curve is 0.191; and this is comfortably close to the value 0.189, which is the value as calculated from the accurate binomial formula.

What Is Normally Distributed?

In a great many circumstances if one obtains a set of numbers which measure some characteristic of the objects in question, it turns out that these numbers are "normally" distributed to some useful approximation. That is, if you make an appropriate chart out of the numbers, you get something that looks very like the Gaussian curve.

Two cautions should be expressed at once. First, you must not think that a quantity is *abnormally* distributed (using the word "abnormal" with its usual bad implications) just because it may not be *normally* distributed. And second, there are in fact a good number of circumstances leading to distribution charts which are *not* normal or Gaussian.

Some things tend to shift unstably away from a medium size, and to be either pretty small or pretty large. This leads to a distribution chart of a very different sort.

For example, consider the fraction of the sky covered, at any one moment, by clouds. There is a strong tendency toward the two extreme cases of zero and one. That is, the sky tends to be either very cloudy, or very clear; and on the *fewest* occasions do you find just half clear and half cloud. This leads to a U-shaped distribution curve; with maxima on the ends, and the minimum in the middle — precisely *unlike* the Gaussian case.

Then suppose, for a large number of families, you plot the distribution of incomes. Most families have small or modest incomes, and as the curve moves to the right toward larger and larger incomes, the number of families with those incomes becomes smaller and smaller. This leads to a distribution curve which is very roughly J-shaped.

There are also various distributions which, looking roughly like the Gaussian one, are not symmetrical about the mid-point, but are *skewed.*

Thus here is a record of weights of a large group of men, all between 25 and 30 years of age, and all within a half-inch of 5′6″ in height:

Weight in pounds	105	120	135	150	165	180	195	210
No. of cases	17	722	2175	1346	483	155	33	3

The average weight of these men is approximately 141.6 lb. But if you plot a chart, then it is clear that the distribution is "skewed" over toward the lesser weights (Fig. 38).

Again, if you count the frequency of the letters of the alphabet in a long piece of ordinary prose,[3] then you find that the letter *E* occurs about 12.3 per cent

[3]Not, for example, Lewis Carroll's description of the game he invented and called "Syzygies and Lanricks."

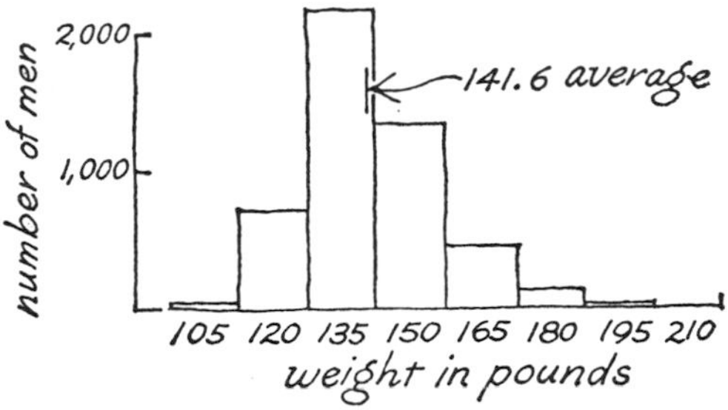

FIG. 38 *Distribution of men's weights*

of the time, T about 9.6 per cent, A about 8.0 per cent and so on, down to Q, X, J, and Z, which are at the end of the list. Since we now know that the observed frequencies in a large sample are good guides to the probability of encountering the letters in picking out a letter by random choice (say by closing your eyes, and making a dot on a page of a book with a pencil, then taking the letter nearest the dot), you could, from the tables of letter-frequencies used in cryptanalysis, make a probability distribution chart. If you put the letters down in their natural order, the resulting distribution chart would be a very bumpy affair, as shown in Fig. 39.

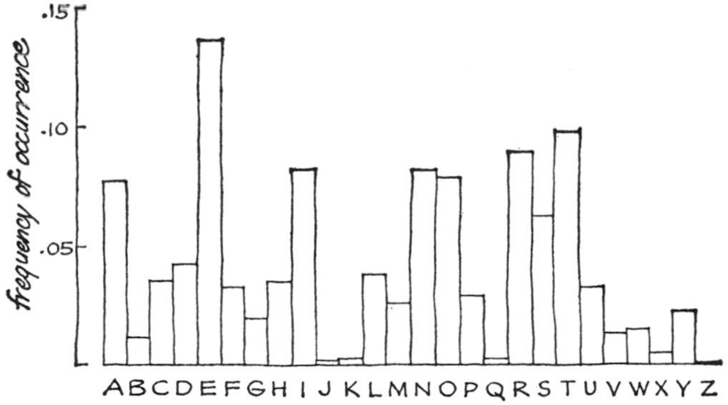

FIG. 39 *Frequencies of occurrence of letters*

Since, however, the arrangement of the horizontal scale here is so arbitrary, we can rearrange the letters and get a chart like Fig. 40, and this does look rather like a flattish Gaussian chart.

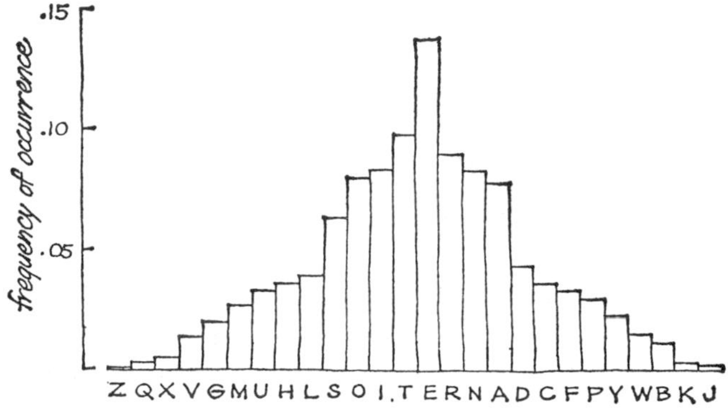

FIG. 40 *Frequencies of occurrence of letters when rearranged*

To return to the title of this section, a great many things *are* normally distributed. The record of successes in any reasonably long binomial experiment gives a normal distribution, as we have seen. Recall that the phrase "binomial experiment" covers any situation in which there are repeated trials (under constant conditions assuring that the probability does not change from trial to trial) which can result in success or failure, and you then begin to realize what a very broad spread of experience is involved, all of which leads, with greater or less precision, to a Gaussian distribution curve.

If you plot a frequency chart of the heights of a considerable group of persons, say 1078, then you get a chart like Fig. 41.[4]

[4]These data come from the classic study "On the Laws of Inheritance in Man," by Karl Pearson and Alice Lee. *Biometrika*, Vol. II, No. 4, November 1903, page 96.

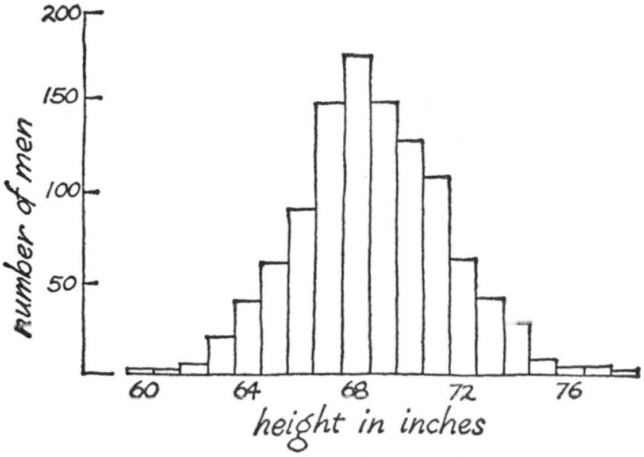

FIG. 41 *Distribution of men's heights*

If you carry out a number of careful measurements of any quantity and plot a chart showing the frequency with which individual measurements differ from the average of all the measurements, then this "error curve" turns out to be shaped like the Gaussian curve. We will return to this topic in Chapter XIV, to indicate how important are the consequences of this fact.

We have spoken of two very wide categories of instances to which the normal distribution applies — the results of any binomial experiment and of any extensive measuring process. Closely related with the second of these is still another very broad category of cases. When you take the sum of a number of random variables, then whether or not they are normally distributed individually, it turns out that the *sum* is normally distributed. This is a most powerful and useful fact. You may be concerned with a number of quantities whose means and whose standard deviations you know; and you can then make probability calculations concerning their sum, using results from the Gaussian curve, even though you know nothing about the probability distributions of the individual quantities.

That this third category of applications may be closely related to the second category (measurements) is clear if you consider that in many cases the error in a measurement may be the sum-effect of a lot of little individual contributory errors, whose separate distributions you do not know or need to know.

There are many, many more sets of numbers which, when plotted out in a frequency chart, have approximately normal distributions. The results when a large group of high school boys and girls take IQ tests, or College Board aptitude or achievement tests, give curves with this bell-shaped appearance. The deviations from the point of aim in artillery fire and bombing follow the Gaussian law. The velocities of the individual molecules forming a gas are distributed this way, but skewed. The dimensions of all sorts of naturally occurring objects — for example, the lengths of the leaves on a tobacco plant — conform to the normal or Gaussian distribution.

The Quincunx

In 1889 Sir Francis Galton, the English expert on heredity, wrote a famous book called *Natural Inheritance;* and in this book he described what he called a Quincunx.

Suppose you have, fastened in a large board, a number of round pegs, each say one-quarter of an inch long, and arranged in equally spaced rows, the pegs in any one row being midway between the pegs in the row just above, as shown in Fig. 42.

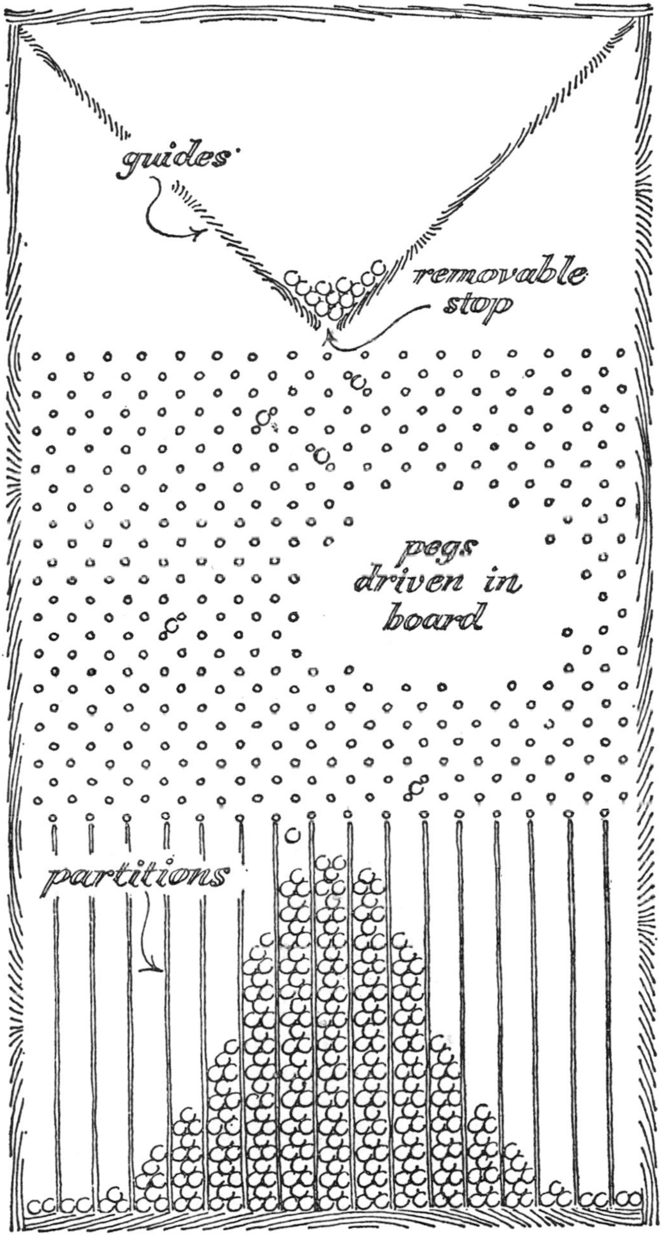

FIG. 42 *The Quincunx*

Now fasten a sheet of glass over all the pegs, stand the board up vertically, and arrange a receptacle at the top (Fig. 42) into which you can put a large number of small round balls (say ball bearings) considerably smaller, in their diameter, than the spacing between pegs. Fix up slanting guides to bring the balls to a center exit, and a stop to keep them in the upper receptacle until you are ready for the show to start. The little gate through which the balls can fall, when the exit stop is removed, is centered carefully above the middle peg in the top line of pegs. The glass plate serves to keep the balls from bouncing out, and also permits you to see what goes on.

A ball falls on the very top of the middle round peg. It is as likely to fall to the right as to the left. If it moves to, say, the left, it finds itself in a new "gate" centered on a peg in the second line; so that in its second bounce it is again as likely to move one space to the left as to move one space to the right.

Now recall the arithmetic triangle of Pascal, which we saw back on page 213.

1
1 1
1 2 1
1 3 3 1
1 4 6 4 1
1 5 10 10 5 1
1 6 15 20 15 6 1
1 7 21 35 35 21 7 1
1 8 28 56 70 56 28 8 1

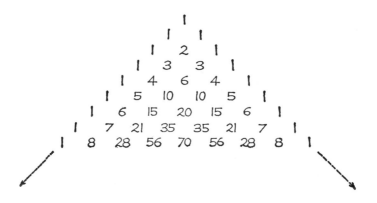

It was stated that the simple process of adding two adjacent numbers in one row produces the correct number in the next lower row midway between the two added. We know that this procedure produces the binomial coefficients.[5] Furthermore we know, from equation (23), that

$$\binom{N}{m}p^m q^{N-m} = \binom{N}{m}\frac{1}{2^N}$$

is the probability for m "successes" and $N - m$ "failures" in a series of N trials in each of which $p = q = 1/2$. Furthermore, we know that a *big* binomial experiment gives a result very like a Gaussian distribution.

So if you pull out the stop and let all the balls trickle down, what will the end result look like if you have arranged little vertical compartments at the bottom to catch and retain the balls?

You get, of course, a nice approximation[6] to the Gaussian distribution curve.

It remains only to explain why Sir Francis called this a Quincunx. There was a Roman coin of that name which was worth five-twelfths of a lira, and on

[5]You may want to *prove* this. Two adjacent binomial coefficients in the Nth row — say the $(r - 1)$'st one and the r'th one, are $\binom{N}{r-1}$ and $\binom{N}{r}$ respectively. Add these and you do, after a little bit of very easy algebra, get $\binom{N+1}{r}$.

[6]The Science Materials Center, 59 Fourth Avenue, New York 3, N.Y. has a very clever, entertaining, and instructive probability and statistics "kit" (with an instruction book by Edmund C. Berkeley and a lot of experiments to do) which includes a device equivalent to the Quincunx.

the face of which were five spots arranged in the pattern

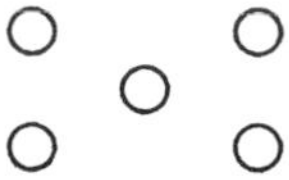

Other Probability Distributions, The Poisson Distribution

Although the Gaussian probability distribution is by all odds the most important one, with the largest range of application, there are various other probability distributions which have been found useful. Of these we will take a quick look at only one — the so-called Poisson distribution.

The distribution did not arise in connection with some piscatorial problem, but was first introduced by Siméon Denis Poisson in a book he wrote a century and a quarter ago dealing with the applications of probability theory to lawsuits, criminal trials, etc.

The Poisson distribution can be derived from, and as a special case of, the Gaussian distribution. For the Poisson distribution applies when the probability p for "success" on any one trial is *very small*, but the number of trials N is *so large* that the expected number of successes, pN, is a moderate-size quantity. Thus pN could have the value 2 in the case of 2000 trials in which the probability of success on one trial is only 1/1000. Since, in discussion of the Poisson case, the expected value pN is specially important, let's give it a special notation by setting $pN = a$.

The smallness of p forces the total unit area under the distribution curve to be concentrated in the neighborhood of $m = a$. And since only an integral number of successes can occur, most of the unit supply of probability has to be assigned to a few integers near $m = a$.

In these circumstances the probability of exactly m successes in N trials can be shown to be

$$W_m = \frac{a^m}{m!} e^{-a} \tag{33}$$

where I have used the new notation W_m for m successes, this notation to be reserved for the Poisson case. This expression is completely consistent with our earlier general expression (29) for m successes in N trials, and also with the value that would be obtained from the Gaussian curve [equation (32)]. It is just a more simple and convenient form to use when p is small, N large, and $a = pN$ moderate in size.

A classical case to which the Poisson distribution was applied is that of the deaths which occurred over the years 1875 to 1894 in various corps of the German army, these deaths being of troopers kicked by cavalry horses. The actual data looked like this:

Summarizing the raw data,

Deaths per Year	*No. of Cases*
0	144
1	91
2	32
3	11
4	2
5 or more	0

Since a total of 280 "experiments" are listed (number of years times number of corps), and since there were 196 deaths, the average expected number

TABLE XII
MEN KILLED BY BEING KICKED BY CAVALRY HORSES IN THE GERMAN ARMY

Year	1875					1880										1890				1894
Guard Corps G	0	2	2	1	0	0	1	1	0	3	0	2	1	0	0	1	0	1	0	1
Army Corps 1	0	0	0	2	0	3	0	2	0	0	0	1	1	1	0	2	0	3	1	0
Army Corps 2	0	0	0	2	0	2	0	0	1	1	0	0	2	1	1	0	0	2	0	0
Army Corps 3	0	0	0	1	1	1	2	0	2	0	0	0	1	0	1	2	1	0	0	0
Army Corps 4	0	1	0	1	1	1	1	0	0	0	0	1	0	0	0	0	1	1	0	0
Army Corps 5	0	0	0	0	2	1	0	0	1	0	0	1	0	1	1	1	1	1	1	0
Army Corps 6	0	0	1	0	2	0	0	1	2	0	1	1	3	1	1	1	0	3	0	0
Army Corps 7	1	0	1	0	0	0	1	0	1	1	0	0	2	0	0	2	1	0	2	0
Army Corps 8	1	0	0	0	1	0	0	1	0	0	0	0	1	0	0	0	1	1	0	1
Army Corps 9	0	0	0	0	0	2	1	1	1	0	2	1	1	0	1	2	0	1	0	0
Army Corps 10	0	0	1	1	0	1	0	2	0	2	0	0	0	0	2	1	3	0	1	1
Army Corps 11	0	0	0	0	2	4	0	1	3	0	1	1	1	1	2	1	3	1	3	1
Army Corps 14	1	1	2	1	1	3	0	4	0	1	0	3	2	1	0	2	1	1	0	0
Army Corps 15	0	1	0	0	0	0	0	1	0	1	1	0	0	0	2	2	0	0	0	0

a (of deaths per experiment) is 0.7. Thus the Poisson law for this situation is

$$W_m = \frac{(0.7)^m}{m!} e^{-0.7}$$

and from this one can very easily calculate the theoretical number of times that the deaths per year, in 280 experiments, should be 0, 1, 2, etc.

For example

$$W_1 = \frac{0.7}{1} \cdot e^{-0.7} = 0.3476$$

so that in 280 "experiments" one could reasonably expect one death to occur 280 (.3476) = 97.3 times. This did occur 91 times.

The comparison between theory and experience, for these cavalry deaths, is certainly impressive. Here it is:

No. of Deaths per year	*Actual Number of Instances*	*Theoretical Number of Instances*
0	144	139.0
1	91	97.3
2	32	34.1
3	11	8.0
4	2	1.4
5 or more	0	0.2

One of the considerable advantages of the Poisson distribution law is that one does not need to know the value of the small probability p, nor of the large number of trials N. He only needs to know the quantity $a = pN$.

Thus suppose a small grocery store sells, on the average, 2 jars of caviar per week, and brings their stock up to 4 jars every Monday morning when the wholesale delivery truck calls.

What, in these circumstances, is the chance that the owner will sell only one jar, or none at all, or that there will be a concentration of demand that cleans out his stock and produces an unsatisfied customer? We do not need to know how many customers N are tempted to be extravagant, nor do we know the level p to which they are tempted. We only need to know that $a = pN = 2$.

The formula for his case being

$$W_m = \frac{2^m}{m!} e^{-2}$$

it is very easy to calculate that

W_m	m
0	.135
1	.276
2	.276
3	.181
4	.092
5	.036

Notice that, 2 being his average weekly sale, the probability of 1 sale a week is just as likely as 2; and that the probability of 5 sales a week is only 0.036 or about 4 in 100.

If he sells exactly 4 jars a week, he does not sell exactly any other number, so these different sales are mutually exclusive events. If you add up the probabilities just given, you get 0.983; which shows that practically all the probability is used up on the cases listed, and that it is *extremely* unlikely that the store will sell more than 5 jars in any one week.

From the Poisson law, the probability that the number of successes will be s or more is

$$P = 1 - e^{-2}\left[1 + a + \frac{a^2}{2!} + \frac{a^3}{3!} + \cdots + \frac{a^{s-1}}{(s-1)!}\right]$$

This expression is useful in so many practical circumstances that it has been given a name (the "Poisson exponential summation") and tabulated.[7] You may be curious to know why this kind of a question comes up in the telephone business; but if you think a moment you will realize (just as one example) that the probability that any one person will call your phone within any one hour is very small indeed; but there are many persons who *might* call. Thus a whole range of problems about frequency of calls, chance of busy signals, equipment necessary to handle traffic, etc., are calculable by probability theory and often by using the Poisson approximation.[8]

The Poisson exponential summation can be used to answer at once a stock inventory question similar to the caviar problem. If a store sells on the average 10 packages a week of a certain item, and each Monday stocks 10, 11, 12 . . . packages, what is the probability that the demand will exceed the supply?

Supply	*Probability Demand Will Exceed the Supply*
10	0.41
11	.30
12	.20
13	.14
14	.08
15	.05
16	.02
17	.015
18	.007
19	.003

[7]F. Thorndike, *Bell System Technical Journal,* November 1926.

[8]See *Probability and Its Engineering Uses,* Thornton C. Fry (New York: D. Van Nostrand Company, 1928).

You can easily see that by taking into account perishability, cost of shelf and storage space, negative value of disappointed customers, etc., the storekeeper can work out a really rational basis for his inventory.

Because the Poisson law applies so neatly to the sales of caviar or dog biscuits, do not make the error of supposing that it does not also bear on questions of greater importance. The disintegration of a radioactive substance follows this law. The number of V-2 bomb hits experienced during World War II by 576 areas of equal size in the south of London followed this law accurately.[9] The chromosome interchanges within cells, bacterial and blood counts, and indeed the number of wars per year (as was shown by the meteorologist Lewis Fry Richardson) and the number of strikes in British industry per week (as was shown by Maurice Kendall of the London School of Economics) all follow the Poisson law.

The Distribution of First Significant Digits

I have been told that an engineer at the General Electric Company, some twenty-five years or so ago, was walking back to his office with a book containing a large table of logarithms. He was holding it at his side, spine down; and as he glanced down at the edges of the pages, he noticed that the book was dirtiest at the opening pages and became progressively cleaner — just as though the early parts of the book had been consulted a lot, the middle less, and

[9] "An Application of the Poisson Distribution," by R. D. Clarke. *Journal of the Institute of Actuaries*, Vol. 72 (1946), p. 48.

the concluding part least of all. "But that," he must have thought, "is ridiculous. That implies that people most frequently look up the logarithms of numbers beginning with the digit 1, next most frequently numbers beginning with 2, and so on, and least frequently numbers beginning with 9. And this just can't be so; because people look up the logarithms of all sorts of numbers, so that the various digits ought to be equally well represented."

Suppose you offer a friend a bet. You tell him that he can open at random any volume (such as the *World Almanac*) which contains tables of all sorts of numbers — the populations of the counties of the U. S., or the areas of seas, or the highest and lowest altitudes in each state, or the students enrolled in all the colleges and universities, or the motor vehicle deaths by states, or whatever. From such a table you and he agree to pick out twenty numbers — the first twenty, or every other one until you have twenty — or according to any other scheme. You say: "Now you are my friend, and I want to treat you right. There are *nine* integers 1 through 9, and nine can't be divided evenly. So I will give you a generous proposition. For every number that begins with 1, 2, 3, or 4, you pay me \$1. For every number that begins with 5, 6, 7, 8, or 9 I will pay you \$1. I pay on five numbers and you pay on only four, which just proves how generous I am."

What happens? Just as an illustration, on page 297 of the 1962 *World Almanac* (which page I just turned to by chance) there is a list of the populations of "Standard Metropolitan Statistical Areas," starting with Abilene, Texas (120,377) and ending, on that page, with Lancaster, Pa. (278,359). There are 92 entries in all, and of these just 20 begin with the digits 5, 6, 7, 8, or 9; whereas 72 begin with the digits 1, 2, 3, or 4. The bet pays off $52 on this particular column!

It is really surprising, when you first meet the fact, that the first digits of random numbers are not distributed equally. It is even more remarkable that this fact was apparently discovered so recently.[10] The proportion of numbers beginning with n or less is *not* $n/9$, but is approximately $\log_{10}(n + 1)$. Thus the proportion beginning with 4 or less is $\log_{10} 5 = 0.699$ so that the bet described above has odds in its favor of just about 7 to 3. In the example of the last paragraph, the fraction beginning with 4 or less was $72/92 = .78$, so my chance illustration was in fact a little too good.

To prove the statement concerning $\log_{10}(n + 1)$ is beyond our scope here, but a very rough argument will, I think, make the result less mysterious.

Suppose we have a lot of identical little cards. On one of these cards write the integer 1, on one 2, on one 3, and so on up to 9 on the ninth card. Put these in a hat and stir them up. If we are going to pick out a card at random and play our game, what are my odds for winning?

Clearly they are 4 to 5; for of these nine cards, four pay me, and the other five pay my friend. But now put in the hat one more card, marked 10, and one marked 11, and so on. All these are "my" cards, and as we thus increase the number of cards in the hat the

[10]In his paper "On the Distribution of First Significant Digits" (*Annals of Mathematical Statistics*, Vol. 32, No. 4, December 1961), Roger S. Pinkham says that there have been only five papers published on this subject, the first in 1938.

game keeps getting more and more favorable to me until we reach a card marked 49. At that moment there are in the hat 44 cards that pay me, and only the original 5, 6, 7, 8, and 9 that pay my friend. In other words, at this juncture the odds in my favor are 44 to 5, or nearly 8.8 to 1.

If we now proceed to number more cards and put them in the hat, all those from 50 to 99 are my friend's cards. When there are cards that exhaust the integers up to 100, the odds for me have been reduced back to 4 to 5 against me.

But as we keep putting more and more cards in the hat, numbered consecutively, my odds keep going up again, and reach approximately 8.1 to 1 at 499, when my friend has 55 favorable to him, and I have 444 favorable to me. Then again, as we proceed, my odds keep dropping until there are 999 cards in the hat, when the odds have dropped to 4 to 5 against me. Again the odds rise, this time to 8.0: and so on.

If we make a graph of my probability of winning, for various numbers of cards in the hat, we get a saw-tooth curve with the bottom of the teeth located at 9, 99, 999, 9999, etc. (at which points my odds are 4 to 5, or my probability of winning is 4/9 or 0.444), and with the top of the teeth located at 49, 499, 4999, etc. (at which points my odds are 8.8 to 1, 8.1 to 1, and from there on sensibly equal to 8 to 1, so that my probability of winning is about 8/9 or 0.888.

The breadth of the teeth increase by a factor of ten as one goes from one tooth to the next one to the right. If we made an ordinary picture of this saw-tooth curve, to include say the four teeth from 10 to 100,000, the first tooth would extend only from 10 to 100, the next from 100 to 1000, and the third from

1000 to 10,000: so that the fourth tooth would occupy most of the picture.

We will therefore draw a somewhat distorted picture (see Fig. 43) in which the horizontal scale is non-uniform, but in which the shape of the teeth is the shape they would have (namely straight trailing edges and very slightly curved leading edges[11]) if the horizontal scale were uniform. Don't worry about this little technical complication. The picture will only be used as a general guide to our reasoning.

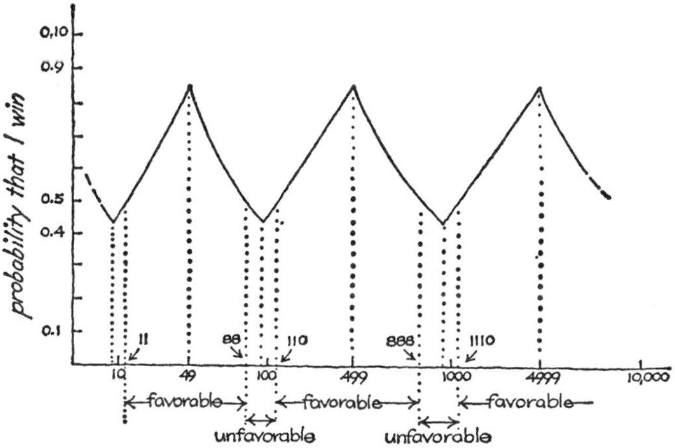

FIG. 43 *Variation of odds with an increasing number of cards in the hat*

Notice on the figure that the probabilities are favorable to me (greater than 0.5) in certain ranges on the horizontal axis; and unfavorable to me (less

[11]In the fractions which represent the probabilities, the increases occur in the numerator on the trailing edges, giving a linear function; but occur in the denominator on the leading edges, giving a hyperbolic function.

than 0.5) for short intervals which occur around the "natural division points" in the decimal number system — that is to say, around 100, 1000, 10,000, and so on.

Now suppose we had a whole lot of hats, with different numbers of cards in them; and we play the game by first choosing a hat at random, and then pulling out cards at random, I winning on any card whose number begins with 1, 2, 3, or 4, and losing on any card whose number begins with 5, 6, 7, 8, or 9.

My probability of winning in this game would be, on the average, the average height of the saw-tooth curve; for picking a random hat is equivalent to picking a random integer on the horizontal axis.

We can estimate the average height of this saw-tooth curve by a simple geometrical trick. The average is, in fact, 0.5 plus the average height of the "favorable triangles" and minus the average height of the unfavorable triangles, weighting the pluses and minuses in proportion to the width of the bases to which they apply.

Using the figures which are rapidly approached as we move out the curve to the right, 22.22% of the numbers correspond to unfavorable mixtures for which the average probability is to be found by subtracting from 0.5 the average height of the little triangles which fall below the 0.5 probability line. The average height of a triangle is half its altitude, and the altitudes of these little triangles have lengths of 0.500 minus 0.444 or 0.055; so that their average height is one half of 0.055 or 0.028.

The triangles which are above the 0.5 probability line correspond to the remaining 77.78% of the numbers, and have altitudes of 0.888 minus 0.500 or

0.388; so that their average height is one half of 0.388 or 0.194.

Thus the average height of the saw-tooth curve is 0.500 plus 77.78% of 0.194 minus 22.22% of 0.028. This gives an average probability, for our numbers game, of 0.645, or odds in my favor of about 6.4 to 3.4.

No one would claim that playing this game with cards in a hat (which amounts to picking at random an integer, and then playing the game with all integers up to and including the one chosen) is precisely the same thing as playing the game with the first digits of random numbers. For that matter, no one should claim that playing the game by taking numbers out of the *World Almanac* is, rigorously, equivalent to playing with truly random numbers.

But the theoretical probability for the first digits with random numbers is 0.699 (or odds of about 7 to 3) whereas we have approximately estimated the probability in our cards-in-the-hat game to be 0.645 (or odds of about 6.4 to 3.4). The two results are not far apart.

The real point in going through these details, however, is to have you get over your surprise that 1, 2, 3, and 4 occur much more frequently, as the first digit of a number, than do 5, 6, 7, 8, or 9. This fact, at first so strange that I have known very able mathematicians initially to pooh-pooh the idea, is simply an inherent characteristic of our decimal number system, as is clearly indicated by the "favorable" and "unfavorable" intervals shown in Fig. 43.

The hat game also makes the following clear. If you try the original bet on a friend, and he becomes too discouraged at his rate of loss, then you can generously offer to shift over to the Manhattan, New York Telephone Directory (using only the last four digits of a telephone number, and throwing out any numbers that begin with zero) and offer to pay *him* on 1, 2, 3, 4, if he pays *you* on 5, 6, 7, 8, 9. The odds now are still in your favor, 5 to 4, for you are now working on one of the bottom points of the saw-tooth curve.

This game, which most of my friends find both interesting and incredible, can be played with almost any set of numbers you can imagine. Pinkham remarks that approximately 0.7 of the physical constants in the Chemical Rubber Company tables "begin with 4 or less," as compared with the theoretical value 0.699. Although it remained unsuspected or at least unidentified for centuries, this distribution law for first integers is a built-in characteristic of our number system. Indeed Pinkham proves, in his article, that the $log_{10}\ (n + 1)$ distribution is the *only* one which retains its essential nature (or as mathematicians say, is invariant) under a change of scale.

The *second* digit of a random number, incidentally, is nearly random, but does show some slight influence of the logarithmic distribution that favors the lower integers. By the time you advance to the third integer, the logarithmic effect is essentially gone.

Chapter XIII

RARE EVENTS, COINCIDENCES, AND SURPRISING OCCURRENCES

"There's no use trying," Alice laughed, "one can't *believe impossible things."*

"I daresay you haven't had much practice," said the (White) Queen. "When I was your age I always did it for half an hour a day. Why, sometimes I've believed as many as six impossible things before breakfast."

LEWIS CARROLL, *Alice in Wonderland*

Coincidences, in general, are great stumbling blocks in the way of that class of thinkers who have been educated to know nothing of the theory of probabilities — that theory to which the most glorious objects of human research are indebted for the most glorious of illustrations.

Arsene Dupin, in "*The Murders in the Rue Morgue,*" BY EDGAR ALLAN POE

Lest men suspect your tale untrue,
Keep probability in view.

JOHN GAY

Well, What Do You Think about That!

I am writing the first draft of this chapter in Naples, Italy, far along on a round-the-world trip that started in the Orient. In the airport at Hong Kong, having a half hour or so to wait, I ran into Professor Joseph Kaplan, a well-known meteorologist and geophysicist of the University of California at Los Angeles, and his wife. They were on their way home from an assignment in West Pakistan. While speaking to them I caught sight, across the room, of Dr. Fred I. Soper, an old Rockefeller Foundation colleague of mine, whom I had not seen for many years. He was returning from a cholera assignment in the Near East. Five minutes after that another man came up and greeted me. It was Dr. Walter S. Salant, of the Brookings Institution, just returning to the U. S. from Indonesia.

Two weeks ago in the American Express office in Florence my eyes fell on Dr. Robert M. Stecher, of Cleveland, a distinguished expert on rheumatism and arthritis. He and I serve together on the Board of Trustees of the Sloan-Kettering Institute for Cancer Research, in New York City. A day or two later, coming out of the Uffizi Gallery, I ran into Wolf Ladejinsky, a friend of mine, who less than ten days ago had resigned his long-term assignment in Viet-nam.

Five human paths, four of them originating in far places, crossed my personal path at the moment I was at the crossing point.

Admitting that ours is, or rather, has become a small world, and admitting that I travel a fair amount and that my business puts me in touch with persons who also travel, it nevertheless gives one something of a start when a series of such events occurs. I should perhaps add that I met, in a hotel in New Delhi, a Dr. Weber who explained that part of his family were named Weaver, since a greatgrand-parent in Ohio had translated his name from German to English. The point is that *my* great-grandfather Weber, in Ohio, changed *his* name (and hence mine) from Weber to Weaver.

Are these things strange? Are they surprising? Does the theory of probability say they shouldn't or couldn't occur?

Before we start a mathematical discussion of such situations, suppose I list a few more instances. I will have to use great restraint, for I have been collecting items of this sort for years.

(A) *Life* Magazine[1] reported that all fifteen members of the choir of a church in Beatrice, Nebraska, due at choir practice at 7:20 P.M., were late the evening of March 1, 1950. The minister and his wife and daughter had one reason (his wife delayed to iron the daughter's dress); one girl waited to finish a geometry problem; one couldn't start her car; another couldn't start her car; two lingered to hear the end of an especially exciting radio program; one mother and daughter were late because the mother had to call the daughter twice to wake her from a nap, and so on. The reasons seemed rather ordinary, but there were ten separate and quite unconnected reasons for the lateness of the fifteen persons.

It was rather fortunate that none of the fifteen arrived on time at 7:20 P.M., for at 7:25 P.M. the church building was destroyed in an explosion. The members of the choir, *Life* reported, wondered if their delay was "an act of God."

[1]*Life*, March 27, 1950, p. 19.

(B) The New York *Herald Tribune* of June 11, 1950, reported that an unidentified man walked up to a dice table at Las Vegas and "made an amazing twenty-eight consecutive passes[2] with the dice — and left a hero to the side bettors, who won, the house said, $150,000. Because he was cautious, the man who threw the dice won only $750."

"The Desert Inn," the report continued, "would, no doubt, have gone broke somewhere along the line, if the man had re-bet his winnings every roll. Presuming, that is, he was psychic enough to quit after twenty-eight throws. He lost on the twenty-ninth. Actually, after each win he stuffed a few bills in his pocket, never risking more than $50.

"He monopolized the dice for one hour and twenty minutes in making his point twenty-eight times.

"Other gamblers, packed four deep around the board, were less conservative. Witnesses said Zeppo Marx raked in $28,000. Gus Greenbaum, owner of a rival club, walked out $48,000 to the good. A professional gambler harvested $9000.

"One man, unable to get close enough to bet, offered $500 for a place at the table.

"The club said the chances of making twenty-eight straight passes are about 10,000,000 to 1."

(C) I believe it authentically reported, although I cannot produce really good evidence, that "even"

[2]In craps a player rolls two dice. If he gets either a 7 or an 11 he wins at once whatever has been bet. If he throws 2, 3, or 12 he loses at once whatever has been bet. If he initially throws 4, 5, 6, 8, 9, or 10 he keeps on rolling the dice until either he "makes his point" by repeating the number he rolled initially, or "craps out" by rolling a 7. If he "makes his point" he wins the bet, and if he "craps out," with a 7, he loses it.

came up once at Monte Carlo twenty-eight times in succession. At Monte Carlo the roulette wheel carries numbers 1 to 36 inclusive, and also the number 0, which is ruled to be neither odd nor even. Thus the chance that the ball will come to rest on an even number is 18/37, or only slightly under one-half. In other words, "even" twenty-eight times in a row is very much the same as heads twenty-eight times in succession when tossing a coin. The probability of this is 1 divided by 268,435,456. Since a table at Monte Carlo makes nearly 500 coups per day, with three different and independent plays for "even" at each table at each coup, it turns out that about every 500 years, on the average, an uninterrupted run of twenty-eight "evens" should occur. There are some four or five tables in use, on the average, at Monte Carlo, so the fact that such a run *did* occur in the first seventy years or so of play there is not, after all, very surprising.

(D) My next-door neighbor, Mr. George D. Bryson, was making a business trip some years ago from St. Louis to New York. Since this involved weekend travel and he was in no hurry, since he had never been in Louisville, Kentucky, since he was interested in seeing the town, snd since his train went through Louisville, he asked the conductor, after he had boarded the train, whether he might have a stopover at Louisville.

This was possible, and on arrival at Louisville he inquired at the station for the leading hotel. He accordingly went to the Brown Hotel and registered. And then, just as a lark, he stepped up to the mail desk and asked if there was any mail for him.

The girl calmly handed him a letter addressed to "Mr. George D. Bryson, Room 307," that being the number of the room to which he had just been assigned.

It turned out that the preceding resident of Room 307 was another George D. Bryson, who was associated with an insurance company in Montreal but came originally (see the next paragraph) from North Carolina. The two Mr. Brysons eventually met, so each could pinch the other to be sure he was real.

There is a sequel. My friend George D. Bryson (the New York City — Connecticut one) tells me, "Later I discovered that my father's father (my grandfather) had gone into the Civil War from a little town in Indiana, just before his son George (my father) was born. When the war was over he was reported missing. Later, he returned to Cincinnati and died there and at that time showed a relative from his hometown a picture of a second family which he said he had in North Carolina where he found himself after the war. It could be that the first-born of that family was also named George, and that just as my father had named me after himself, this other George in North Carolina named his son after himself. That would have made us both grandsons of the same man, but his father illegitimate. Hence, I never explored the possibility with the other George D. Bryson, although I did continue to communicate with him for some years after the coincidence."

(E) During World War II there were two men (the New York *Times* reported this) who, never having previously met, joined up at the same time at Washington, D.C. They received the same assignment, and were together throughout the war. They had the same rank, were of the same size, had similar mustaches, were 35 years old, and looked so much alike that they were regularly taken to be identical twins. One was named Baker and the other Cook.

(F) In Tampa, Florida, so the New York *Herald Tribune* reported, a Mr. Earl M. Lofton sank an "ace" on the 119-yard first hole of the Palma Ceia golf course. The next two players of the foursome were so excited that they fluffed their shots. But the fourth player, Gilbert Turner, said that "a little thing like a hole-in-one" wouldn't bother him. It didn't. He got one too.

(G) I once visited the University of the Andes, on the slope of Mount Serrate outside Bogotá, Colombia. At tea with the Dean of the Faculty, his slight accent gave me the hunch that he originally came from southern Germany. In answer to my question he said that he had been trained in Munich, had served as a multi-tongued agent in the French underground, had taught high school in Peekskill, New York, and had been cook at a summer camp in the Adirondacks. Then he asked about me. I told him I lived near a small village, New Milford, in western Connecticut. "Which way from the village?" "To the east, up Second Hill." "Do you," he went on to my amazement, "turn right off #25 at the cemetery? Do you go around the big bend, and take the first dirt road to the right? Do you then turn in, at the mailbox, and go something over a hundred yards?"

It developed that he had known the previous owner of my land, and had been there several times;

he proceeded to describe the prize individual trees! Remember that this was a bearded Bavarian talking Spanish on the side of a mountain in South America.

(H) In the card room of the Quadrangle Club at the University of Chicago, years ago, a hand consisting of thirteen spades was dealt. The celebrated mathematician Leonard Eugene Dickson was one of the players. (Those who know his interest in bridge realize that the probability of his being one of the players was not far below unity.) At the request of his companions, he calculated the probability of this deal. (It is roughly 10^{-11}. See note at the end of this chapter.) A young know-it-all gaily reported at lunch the next day that he had calculated the probability of dealing thirteen spades, and had found that Dickson had made a mistake. Another famous mathematician, Gilbert Bliss, was present; he properly dressed down the youngster by saying, "Knowing that Dickson calculated a probability and got one result, and you had tried to calculate the same probability but got another result, I would conclude that the probability is practical unity that Dickson was right and you are wrong."

(I) When rockets are sent high above the earth, they usually contain various instruments to measure certain physical quantities such as temperature, air density, intensity of cosmic rays, etc. Down at White Sands, New Mexico, in the spring of 1954, they were going to fire Viking No. 10. One of the physicists pleaded that he be allotted two pounds of the payload for some special photographic material he was specially interested in. Those in charge were very reluctant to agree, for they were essentially certain that when the thin-walled nose cone impacted with the earth, this photographic material would be destroyed. Many of the instruments in rockets radio back to earth their measurements, or store them in some nearly indestructible form, but this photographic material had to be recovered intact to be worth anything.

Now listen to the account of what happened, as reported in one of the official journals[3] of the American Institute of Physics:

"On the morning of May 7, 1954, when Viking No. 10 was to be fired, a group of people from Las Cruces decided to go out for a picnic. Ignoring all signs warning that the White Sands area was restricted, they set up a table covered with a large sun umbrella and proceeded to relax. About a half hour after the rocket was launched and the recovery group were seated in jeeps awaiting sounding data on the probable area of impact, a telephone call came in to the blockhouse stating that a flying saucer had landed. This one was real as it destroyed an umbrella and wrecked a picnic table. The dimensions of the conical aluminum object which had arrived from outer space coincided with that of our nose cone. This proved to be the most efficient recovery in the history of rocketry. While sounding techniques 'pinpoint' the impact area to a radius of about one mile, it is a great help to have a group of people waving at you as you enter the circle. Fortunately, no one was hurt as the group was playing baseball several yards away when our instrument section demolished their table. The

[3] *Physics Today*, October 1960, pp. 20–21.

emulsions were retrieved in perfect condition, thanks to the step deceleration afforded by the umbrella."

(J) In 1928 the New York *Times* carried a cabled story from Paris, dated February 21. It reported the fact that six persons had been found guilty of the "accidental" death of a M. Desnoyelles. He had been in a sanitarium, and a medicine had been prescribed for him. But the chief physician gave M. Desnoyelles a presciption really intended for another patient named Desmalles. Second, the chief physician failed to check that the prescription, dictated to a clerk, was written as he had ordered. Third, the interne who filled the prescription confused two drugs, and introduced one of a poisonous nature. Fourth, the order for the prescription was mistakenly written on a slip used for medicines for internal use, rather than externally as the doctor intended. Fifth, the head pharmacist, who was supposed to check all prescriptions, was busy. He left the matter to his assistant, and she neglected to check. Sixth, an assistant "corrected" the error in name, and wrote on the medicine that it was intended for M. Desnoyelles. Seventh, the interne who administered the medicine disregarded the indicated dose, handed the bottle to the patient, and instructed him to "take a good big drink."

(K) A TV news program reported that on Friday, November 3, 1961, two women in the same hospital both had babies. Certainly that is not surprising, but wait! These two women were of the same age, they were both Irish, their husbands were of the same age — and for both families this baby completed a set of ten, five boys and five girls.

You doubtless have your own "unbelievable" happenings which you will want to add to this list. Although one would be hard put to invent a mathematical model closely and obviously similar to each of these occurrences, one can nevertheless estimate roughly the order of magnitude of the probabilities involved.

For example, looking back at (A), suppose you are willing to admit that each of the reasons or excuses for lateness would occur, on the average, about once in every four choir rehearsals. It is hard to believe that these conscientious and careful persons would be late oftener than that.

But then, since the ten separate and independent reasons occurred simultaneously on one and the same evening, the probability of everyone's being late is $(1/4)^{10}$. This number is slightly smaller than 10^{-6}; so on the basis of this rough assumption one would say that there was about one chance in a million that the ten reasons would occur on the same evening.

If all fifteen members of the choir had each had a separate, independent, accidental reason for being late, then the probability would be estimated at $(1/4)^{15}$, which is roughly one-billionth.

On the basis of our out-of-the-air guess of a probability of 1/4 that a reason would develop to delay one individual choir member on a given evening, one could easily calculate the probability that one *specified* member would be late on a given evening, and all the rest on time. That would be

$$\left(\frac{1}{4}\right)\cdot\left(\frac{3}{4}\right)^{14} \text{ which equals about } 0.004$$

On the same guess, the probability that one member, *unspecified*, would be late on a given evening is 15 times the last written probability; for it is the sum of the probabilities of the 15 mutually exclusive cases in which choir member #1 is the only one late, choir member #2 is the only one late . . . choir member #15 is the only one late.

On the same guess, the probability that at least one choir member would be late on a given evening is

$$1 - \left(\frac{3}{4}\right)^{15}$$

which is about 0.99.

The guess which I made, a few paragraphs back, about the probability of one reason or excuse may not seem reasonable to you. In that case, you can now make your own model for this affair, and calculate some probabilities. I think it is obvious that anyone could make more reasonable assumptions about the underlying probabilities if he could get a look at the attendance record for these choir practices. If these records extended over several years, one could make from them some pretty convincing assumptions.

The models for illustrations (*B*) and (*C*) should be pretty obvious, and you could work out the calculation for (C) yourself. The calculation for (B) is not very difficult, but is somewhat complicated and tedious.

Any model for illustration (D) would involve some preliminary questions. Is the surprising thing that the two immediately successive inhabitants of Room 307 of the Brown Hotel in Louisville, Kentucky, on that particular day, were both named George D. Bryson? Or is the surprising thing simply that the two immediately successive inhabitants of *some* hotel room in *some* city on *some* day should have the same name — and that this should be discovered through a spontaneous act which itself is doubtless rare? What sort of guesses would you have to make, or what sort of knowledge would you like to obtain concerning the frequency of first and last names and of middle initials, before you tried to make some estimate of the probability of this occurrence? Would the story be as good if both men had been named John Smith[4]? Does the unpremeditated stopping in Louisville play any role? Would it have been more or less surprising if it had happened at the Ritz in Paris, or at the biggest hotel in the world?

Small Probabilities

Suppose that an event has a small, even a very small, probability of occurrence; and suppose that an experiment or trial is carried out and the event actually occurs. What should be one's reaction?

If the probability is 10^{-6} or one in a million, then from the definition of probability, and with our ideas sharpened up by what we now know about the law of large numbers, we should expect this event to occur just about once in each million trials in a very long series of many million trials. Therefore we are bound to think about this as a rare event. The smaller the probability, the more rare the event.

Indeed, the law of large numbers and Poisson's

[4]The first twenty most numerous surnames in the United States are, in decreasing order, 1 Smith, 2 Johnson, 3 Brown, 4 Williams, 5 Miller, 6 Jones, 7 Davis, 8 Anderson, 9 Wilson, 10 Taylor, 11 Thomas, 12 Moore, 13 White, 14 Martin, 15 Thompson, 16 Jackson, 17 Harris, 18 Lewis, 19 Allen, 20 Nelson.

law furnish us with exactly the set of ideas we need in order to think, in a correct, quantitative way, about the pattern in which rare events can be expected to occur in a very, very long series of trials.

But just because the event is rare, should we perk up or even be astonished when it does occur? Is the occurrence of a very rare event properly to be viewed as very interesting?

To comment on the last question first, *interest* seems usually to be a very subjective response. Some persons are interested in snakes, others in Renaissance art, and others in football. So I guess all one can say about this last question is, O.K. If the occurrence of this rare event interests you, go ahead and be interested. That is your personal privilege.

But if you use the word *interesting* in the universal sense that every person who knows about the occurrence of the event should be interested — and surprised — and even astonished, then it is necessary to consider the matter more carefully.

Suppose one shuffles a pack of cards and deals off a single bridge hand of thirteen cards. The probability, as reckoned before the event, that this hand will contain any thirteen specified cards is, as mentioned earlier, 1 divided by 635,013,559,600. Thus the probability of any one specified set of thirteen cards is, anyone would agree, very small.

When one hand of thirteen cards is dealt in this way, there are, of course, precisely 635,013,559,600 different[5] hands that can appear. All these billions of

[5]Order of appearance of cards in the dealing process is not taken into account: only the final constitution of the hand. Two hands are, of course, "different" if they differ in one or more cards.

hands are, furthermore, equally likely to occur; and one of them is absolutely certain to occur every time a hand is so dealt. Thus, although any one particular hand is an improbable event, and so a rare event, no one particular hand has any right to be called a surprising event. Any hand that occurs is simply one out of a number of exactly equally likely events, some one of which was bound to happen. There is no basis for being astounded at the one that did happen, for it was precisely as likely (or as unlikely, if you will) to have happened as any other particular one.

So we see that an event should not necessarily be viewed as surprising, even though its *a priori* probability is very small indeed. If you ever happen to be present at the bridge table when a hand of thirteen spades is dealt, what you ought to say is this: "My friends, this is an improbable and a rare event, but it is *not* a surprising event. It is, however, an interesting event."

For it remains stubbornly true that, in the circumstances described, we are sometimes interested in the outcome, and sometimes uninterested. It seems clear that this occurs because the hands, although equally likely, are just not equally interesting. Thirteen spades *is* an interesting hand, and indeed a Yarborough, with no card above the nine, is interestingly poor, whereas millions of the hands are dull and wholly uninteresting.

On the other hand, think of a round disk of metal, like a coin except with perfectly plane faces and no milling on the edge. If this coin-disk when tossed should land in an almost perfectly vertical position, with practically no spin, it just could remain standing on its edge. One could easily adjust the thickness and the mode of spinning so that the probability of its coming down and standing on its edge would be, say, one in a billion.

When this metal disk is tossed, there are, then, three possible outcomes. It may come up heads (for which the probability differs from one half by one half of a billionth), or it may come down tails (which is exactly as likely as heads), or it may stand on its edge (for which the probability is one one-billionth).

Either heads or tails would be probable, not rare, uninteresting, and not surprising. Standing on edge would be improbable, rare, interesting, and surprising. But this surely surprising event is more than 635 times as probable as would be an uninteresting hand in our previous illustration. So standing on edge is not surprising just because it has a low probability. Why is it surprising?

It is surprising, of course, not because its probability is small in an absolute sense, but rather because its probability is so small as compared with the probabilities of any of the other possible occurrences. Standing on edge is half a billion times as unlikely as the only other two things that can happen, namely, heads or tails, whereas the dull hand is precisely as likely (or unlikely) as any of the possible alternative outcomes of the dealing process.

Thus one concludes that *probability* and *degree of rarity* are essentially identical concepts, that a *rare* event is *interesting* or not depending on whether you consider it interesting or not,[6] and that an event is *surprising* only providing its probability is very small

[6]It may perfectly well (alas for after-dinner conversation) be interesting to you, say, because it happened to you, and not at all interesting to anyone else.

as compared with the probabilities of the other accessible alternatives. This requires that a *surprising* event be a *rare* event, but it does not at all require that a *rare* event be a *surprising* event.

Years ago I wrote a paper[7] on this subject in which I proposed a "Surprise Index" which furnished a kind of measure of the amount of surprise you are justified in having when a probability event occurs. Indeed, in any experiment one can easily calculate the size that one can reasonably expect in that experiment of the probability of the outcomes. When you toss a coin, you *know* that you are going to experience an event whose probability is 1/2. Similarly, when you roll two dice, you *know* you are going to experience an event whose probability is 1/36. When you deal a hand of 13 cards from a bridge deck, you *know* you are going to experience an event whose probability is less than 10^{-11}. You have no basis whatsoever for being surprised at an event whose probability is less than 10^{-11} *when you are dealing bridge hands;* but you would be justified in fainting away if an event of this size probability occurred in an experiment with three possible outcomes, the other two having probabilities very nearly equal to 1/2.

Indeed, when a probability experiment is carried out, one of the outcomes necessarily occurs; that is, in fact, what we mean by "doing the experiment." This outcome (the one that actually happens) had a certain *a priori* probability p that it would occur; and, since it has occurred, we can say that we have

[7]"Probability, Rarity, Interest, and Surprise," by Warren Weaver. *The Scientific Monthly*, Vol. LXVII, No. 6, December 1948.

realized, in this experiment, a probability p. How much probability can one expect on the average to realize in a given experimental setup? Mathematical expectation answers precisely this kind of question. We must calculate the expected value of the probability. Thus

$$\begin{aligned} E(P) &= P_1 \cdot P_1 + P_2 \cdot P_2 + \cdots + P_n \cdot P_n \\ &= P_1^2 + P_2^2 + \cdots + P_n^2 \end{aligned}$$

is the average amount of probability we can expect to realize per trial of the experiment in question. And the way that the experiment actually comes out is to be considered surprising only if the probability of the outcome which actually occurred is very small as compared with the expectation in probability for the trial.

In tossing a coin, the expectation in the probability is

$$E(P) = \frac{1}{2} \cdot \frac{1}{2} + \frac{1}{2} \cdot \frac{1}{2} = \frac{1}{2}$$

since the probability is 1/2 of obtaining a head for which the probability is 1/2 (thus giving rise to the term $1/2 \cdot 1/2$), and similarly for tails. This equation says that in tossing a coin you can reasonably expect to experience an event whose probability is 1/2, or (putting it another way) that you can reasonably expect to realize a probability of 1/2. This is an understatement, of course, since in this example you are *sure* to get this much probability.

Similarly for rolling two dice, the expectation in the probability is

$$E(P) = \frac{1}{36} \cdot \frac{1}{36} + \frac{1}{36} \cdot \frac{1}{36} + \cdots + \frac{1}{36} \cdot \frac{1}{36}$$

with 36 terms on the right, so that

$$E(P) = \frac{1}{36}$$

Again, you are sure to realize this much probability.

For a bridge hand the expectation in the probability, using the approximate value 10^{-11} for the probability of any one specified hand, is

$$E(P) = 10^{-11} \cdot 10^{-11} + \cdots + 10^{-11} \cdot 10^{-11}$$

there being 10^{11} terms, one for each possible (and equally probable) hand. Thus for this case,

$$E(P) = 10^{-11}$$

and it is true that, when you shuffle and deal a hand, you are bound to experience an event whose probability is exactly equal to the expected probability in the trial.

For a coin, the normal, non-surprising amount of probability is 1/2. For two dice it is 1/36. For a bridge hand it is 10^{-11}.

A hand of thirteen spades is rare and interesting, but should not be viewed as surprising. How about two hands of solid trumps in one evening or, what is entirely equivalent, any two identical hands in one evening? That is a very different matter!

If an experiment consists of dealing just two hands, then this experiment can result in either onc of two outcomes. It may result in two unlike hands, or it may result in two identical hands. One sees at once that the latter of these two outcomes has an *extremely* small probability, and the former a probability very close indeed to 1. Therefore (check this) the expectation in the probability for this experiment is very nearly unity. One has no basis for surprise if the two hands are different (in that case you are experiencing a probability very nearly equal to the expected probability); but one would have every right to be surprised indeed if the two hands were the same (for then you would be experiencing a probability exceedingly small as compared with the expected probability).

This last calculation also indicates that there is in fact a viewpoint which justifies surprise at a single hand of thirteen spades. Suppose, for example, that one is so much interested in a perfect hand that he lumps together all imperfect ones. That is, for this person there are only two sorts of hands, perfect and imperfect. For him the deal of a hand comes out in only two ways. One of these two ways has an exceedingly low probability and the other has a probability exceedingly close to one. So the last sentence of the preceding paragraph applies directly. It is therefore necessary to modify the general statements made previously and say that when an improbable event is so interesting that all its alternatives are lumped together as events so dull as to be indistinguishable, then the interesting event may thereby become a surprising event.

It is a good thing to notice that, no matter how small a probability you may wish to mention, in the hope that you are characterizing something so improbable that "it just couldn't happen," it is simplicity itself to arrange circumstances so that something just as, or more, improbable simply *has* to occur. Perhaps the easiest thing would be to have a pack of M cards, all different. If M is 52, and if you shuffle and deal hands of 13 cards each, the probability for any one specified hand (which as has been emphasized is the probability for each and every other specified hand) is of the order of 10^{-11}. This probability, for a specified hand of 13 cards, decreases, of course, as the number of different cards in the deck is increased. For a deck of 104 cards the probability of a specified hand of 13 cards is roughly 10^{-16}; for a deck of 208 cards it is about 10^{-20}; and for a deck twenty times the normal size (that is, with 1040 cards) the probability for a specified hand of 13 cards is about 10^{-32}.

Thus if you were shuffling a deck of 1040 cards, and then dealing off a hand of 13, the event which occurs has a probability of about 10^{-32} — one in one hundred thousand billion billion billion. And of course, in those circumstances *an* event of that very small probability is *bound* to happen. Remember this the next time someone calculates that the probability of some event is 10^{-32}, and then concludes "therefore it simply could not have happened by chance."

Note on the Probability of Dealing Any Specified Hand of Thirteen Cards

What is the probability that, from a well-shuffled deck of 52 cards, you deal off 13 cards, one after another, and then find that you have thus obtained a "specified hand," i.e., that you have obtained (the order of obtaining them not being important) exactly the group of 13 cards you specified before dealing?

Think of putting a red mark on each of the 13 specified cards, and a black mark on all the 39 others. Then clearly a model for our problem is this: There are 13 red balls and 39 black balls in a box. Mix them up, and draw 13 out at a time without returning any ball. What is the probability that you have drawn the 13 red balls?

The probability that the first ball drawn will be red is 13/52, for there are, for this one drawing, 13 outcomes which lead to a red ball, out of the 52 equally favorable total outcomes. The probability that the second will be red is now 12/51; for there were 12 red balls out of 51 in all. The compound probability that the first will be red and that the second will be red is (13/52)(12/51). Thus, proceeding in this way, the probability that the 13 drawings each will produce a red ball is

$$\frac{13\cdot 12\cdot 11\cdot 10\cdot 9\cdot 8\cdot 7\cdot 6\cdot 5\cdot 4\cdot 3\cdot 2\cdot 1}{52\cdot 51\cdot 50\cdot 49\cdot 48\cdot 47\cdot 46\cdot 45\cdot 44\cdot 43\cdot 42\cdot 41\cdot 40}$$

If you carry out cancellation, this is equal to

$$\frac{1}{49\cdot 47\cdot 46\cdot 43\cdot 41\cdot 17\cdot 10\cdot 5\cdot 4} = \frac{1}{635{,}013{,}559{,}600} = 1.6\times 10^{-12}$$

FIG. 44

If four persons are seated around a bridge table, and the deck is shuffled and dealt out, what is the probability that *some* one of the four players gets a hand which is specified in advance?

Turning to the box of balls, we must draw them all out, without returning, one at a time. For the person to the left of the dealer to obtain the specified hand, the balls must come out R, B, B, B, R, B, B, B, R, B, B, B, . . . the first ball and every fourth ball thereafter being red. The probability of this is

$$\frac{13\cdot 39\cdot 38\cdot 37\cdot 36\cdot 12\cdot 35\cdot 34\cdots 1\cdot 3\cdot 2\cdot 1}{52\cdot 51\cdot 50\cdot 49\cdot 48\cdot 47\cdot 46\cdot 45\cdots 4\cdot 3\cdot 2\cdot 1}$$

$$= \frac{13\cdot 12\cdot 11\cdots 2\cdot 1\cdot 39\cdot 38\cdot 37\cdots 2\cdot 1}{52\cdot 51\cdot 50\cdots \qquad 3\cdot 2\cdot 1}$$

and canceling, it is

$$\frac{13\cdot 12\cdot 11\cdots 2\cdot 1}{52\cdot 51\cdot 50\cdots 41\cdot 40}$$

which is exactly the same as the probability we calculated above. But we must go on and include the instances in which the dealer's partner, the person to the right of the dealer, or the dealer himself is to get the specified hand. Each of these has the probability we just calculated, and the total probability for the four mutually exclusive outcomes is the sum of the four probabilities, which is 4 times one of them.

Thus the probability that a specified hand will appear at some one of the four table positions is

$$\frac{4}{635{,}013{,}559{,}600} = \frac{1}{158{,}753{,}389{,}900} = 6.3 \times 10^{-12}$$

Someone always writes a letter to the paper if a hand of 13 spades is dealt. But suppose one of your bridge companions, just after being dealt a hand some evening, stopped and said, "Fellows, this is just too improbable to let pass unnoticed! I am sorry to spoil the deal, but I just *have* to tell you what just happened. *I have been dealt, in spades, the Queen, Jack, 6, and 3; in hearts the Ace, 10, 8, 5, and 2; in diamonds the King and 7, and in clubs the 9 and 4!* Do you realize how rare and improbable this hand is? Has any one of you *ever* seen it before? Let's write a letter to the local paper."

Any comment?

Further Note on Rare Events

The great mathematician J. E. Littlewood, in his delightful book *A Mathematician's Miscellany,*[8] remarks that something like two million people in England each play an average of 30 bridge hands a week; so that it is not at all surprising that the newspaper report of someone holding 13 of a suit at bridge is something of an annual event.

Littlewood includes in his book a short section on "Coincidences and Improbabilities." You will get something of the flavor of his book from the following quotation.

> I sometimes ask the question: What is the most remarkable coincidence you have experienced, and is it, for the most remarkable one, remarkable? (With a lifetime to choose from, 10^6:1 is a mere trifle.) This is, of course, a subject matter for bores, but I own two, one startling at the moment but debunkable, the

[8]London: Methuen & Co., 1953.

other genuinely remarkable. In the latter, a girl was walking along Walton Street (London) to visit her sister, Florence Rose Dalton, in service at No. 42. She passed No. 40 and arrived at 42, where a Florence Rose Dalton was cook (but absent for a fortnight's holiday deputized for by her sister) but the house was 42 Ovington Square (the exit of the Square narrows to road width), 42 Walton Street being the house next further on. (I was staying at the Ovington Square House and heard of the occurrence the same evening.)

. . . There must exist a collection of well-authenticated coincidences, and I regret that I am not better acquainted with them. Dorothy Sayers in *Unpopular Opinions*, cites the case of two negroes, each named Will West, confined simultaneously in Leavenworth Penitentiary, U.S.A., in 1903, and with the same Bertillon measurements. (Is this really credible?)

Eddington once told me that information about a new (newly visible, not necessarily unknown) comet was received by an observatory in misprinted form; they looked at the place indicated (no doubt sweeping a square degree or so) and saw a new comet. (Entertaining and striking as this is, the adverse chance can hardly be put at more than a few times 10^6).

Littlewood is certainly correct in remarking that this is a subject for bores. In any event, it is very easy to become addicted to the collection of coincidences. Within five minutes after writing the above paragraphs I read in the morning newspaper (New

York *Herald Tribune,* Wednesday, March 14, 1962) an article headed "Window Washer Miracle." It explains that the day before Mr. Warren Rogers was washing the outside of a third-floor window of the Fox Medical Building. The safety hook tore loose, and he found himself dangling by one strap 40 feet above the sidewalk. Managing to pull himself back in he decided the only thing to do was to go back to work. He started to wash the windows on a fifth-floor office, and the safety hook there broke, leaving him hanging 65 feet above the sidewalk.

Chapter XIV

PROBABILITY AND STATISTICS[1]

Statistical thinking will one day be as necessary for efficient citizenship as the ability to read and write.

H. G. WELLS

"Go on, Mrs. Pratt," says Mrs. Sampson, "them ideas is so original and soothing. I think statistics are just as lovely as they can be."

O. HENRY, "The Handbook of Hymen"

Statistics[2]

The word "statistics" has an "s" on the end of it; but it can be used in both a plural sense and in a singular sense.

When used in the plural sense, the word refers to facts which can be stated in numbers or more usually in tables of numbers. Thus one speaks about the

[1]This chapter makes extensive use of a paper I wrote with the title "Statistics" in the *Scientific American*, January 1952, Vol. 186, No. 1, pp. 60–63.

[2]*Statistics,* by W. Allen Wallis and Harry V. Roberts (Glencoe, Illinois: The Free Press, 1956) is a superb general text which makes very modest demands as regards mathematics, and is excellently clear and informative.

statistics of wheat production in the countries of the world, or the statistics of bank clearances.

When used in the singular sense, the same word means something quite different — as Webster says, "the science of the collection and classification of facts on the basis of relative number or occurrence as a ground for induction; systematic compilation of instances for the inference of general truths; the doctrine of frequency distributions."

When one couples the words *probability* and *statistics*, as in the title of this chapter, it is of course the second definition that one has in mind. This second definition may seem a little formidable on first reading; but surely the phrases "relative number or occurrence" and "frequency distributions" give sufficient warning that the theory of probability is not far around the corner.

It may be helpful if, at the very beginning of this chapter, we consider for a moment the relationship between probability and statistics.

It is somewhat tempting to think of probability theory as the "pure" theory, and statistics as the body of information obtained when probability theory is *applied* to a certain kind of practical problem. But this is not a correct or very useful terminology. Mathematical probability is itself concerned with all sorts of applied problems, and statistics has developed a very substantial body of theory more or less peculiar to itself; so the contrast between pure and applied does not successfully serve to differentiate between the two.

It is, however, possible (with a little oversimplification) to indicate a main difference between the two by means of two problems.

If you have a jar inside of which is a number of balls, similar except for color — if you know how many of each color are in the jar — and if you are considering taking out a certain number, keeping the contents thoroughly mixed (returning or not returning after each draw, as may be specified) then *probability theory* (as you know very well if you have stuck with me this far) is prepared to tell you the probability that you will get any sample of stated composition. You know *what's in the jar*, and you ask concerning the probability of compositions of the possible samples you *might draw but have not drawn.* This is a characteristic problem of *probability theory.*

But suppose you do *not* know anything at all about the proportions of colors in the jar, and suppose you *have* drawn out a sample and have observed what it contains — so many white balls, so many black, so many red, etc. Then you can ask, "What is in the jar?" A little more reasonably: "Of various guesses or estimates I may make, based on the evidence of my sample of what is in the jar, which estimate is most probable, and how much confidence can I have that the actual mixture in the jar is in fact this most probable one?" This is a characteristic problem of *statistics.*

If we call the contents of the jar the "population" we are interested in, then we may say, with reasonable accuracy but not with very satisfactory inclusiveness:

> Probability theory computes the probability that *future* (and hence presently unknown) samples out of a *known* population turn out to have stated characteristics.

> Statistics looks at a *present* and hence *known* sample taken out of an *unknown* population, and makes estimates of what the population may be, compares the likelihood of various populations, and tells how confident you have a right to be about these estimates.

Stated still more compactly, probability argues from populations to samples, and statistics argues from samples to populations.

However, statistics, in addition to studying samples and drawing inferences from them, is also very widely and importantly used for a purpose which is, at least superficially, rather different. Namely, if a person is going to do an experiment or a series of experiments (test whether or not a new drug is effective, compare various fertilizers, etc.) then he can go to a mathematical statistician *before* he does any of the experiments, and can ask: "How should I plan and conduct these experiments so that the results, after I get them, will be as useful and illuminating and dependable as possible?"

In other words, mathematical statistics, using at every turn the results and reasoning of probability theory, *draws inferences from samples,* and *gives advance planning for experiments.*

Deduction and Induction

There are two main forms of logical thinking — deduction and induction. For the former we are chiefly indebted to the Greeks, who first saw clearly revealed the great power of announcing general axioms or assumptions and deducing from these a useful array of implied propositions. Inductive thinking, which has been called "the second great stage of intellectual liberation," did not begin to become a systematic tool of man until late in the 18th century. Induction proceeds in the opposite direction from deduction. Starting from the facts of experience, it leads us to infer general conclusions.

Deductive reasoning is definite and absolute. Its specific inferences follow inescapably from the general assumptions. Inductive reasoning, on the other hand, is *uncertain* inference. The concrete and special facts of experience, from which inductive reasoning begins, generally do not lead inexorably to categorical general conclusions. Rather they lead to judgments concerning the plausibility of various general conclusions.

Francis Bacon (1561–1626) was the first properly to emphasize inductive methods as the basis of scientific procedure, but it was not until 1763 that the English clergyman Thomas Bayes gave the first mathematical basis to this branch of logic. To get an idea of what Bayes did, let us look at an admittedly artificial example. Suppose you have a closed box containing a large number of black and white balls. You do not know the proportion of black to white but have reason to think that the odds are two to one that there are about equal numbers of black and white balls. You reach into this box, take out a sample of balls, and find that three-fourths of the sample are black. Now before taking this sample you tended strongly to think that the unknown mixture was half white, half black. After taking the sample you clearly should change your thinking and begin to lean toward the view that black balls outnumber the white in the box. Bayes worked out a theorem which indicates exactly how opinions held before the experiment should be modified by the evidence of the sample.

This theorem is of limited usefulness. It is not capable of creating a judgment out of nothing. It can only take a previous judgment (an *a priori* opinion, as the professionals say) and tell you how you are justified in modifying that opinion on the basis of the new evidence. In very many practical circumstances it is difficult or impossible to start out, at the very beginning, with any reasonable or defensible *a priori* opinion. And in that case, Bayes' theorem cannot build a new judgment, because it has nothing on which to build.

Various theoreticians have tried to invent ways to minimize or even to eliminate this difficulty — or rather this restriction — of Bayes' theorem. And a tremendous amount of very lively debate has resulted. This has led some extremists to throw the theorem entirely away. A more reasonable procedure is to use it with caution, understanding, and restraint.

Thus though the usefulness of this theorem has proved to be limited, it was the beginning of the whole modern theory of statistics, and thus of a mathematical theory of inductive reasoning.

Sampling[3]

The importance of inductive reasoning depends on

[3]*Sampling in a Nutshell,* by Morris James Slonim (New York: Simon and Schuster, 1960) is clear, informative, simple, and amusing.

the basic fact that, apart from trivial exceptions, the events and phenomena of nature are too multiform, too numerous, too extensive, or too inaccessible to permit complete observation. As the author of Ecclesiastes remarked, "No man can find out the work that God maketh from the beginning to the end." (3:11). We can't measure cosmic rays everywhere and all the time. We can't try a new drug on everybody. We can't test every shell or bomb we manufacture — for one thing, there would then be none to use. So we have to content ourselves with samples. The measurements involved in every scientific experiment constitute a sample of that unlimited set of measurements which would result if one performed the same experiment over and over indefinitely. This total set of potential measurements is, as mentioned above, referred to as the *population.* Almost always one is interested in the sample only insofar as it is capable of revealing something about the population from which it came.

The four principal questions to be asked about samples are these: (1) How can one describe the sample usefully and clearly? (2) From the evidence of this sample how does one best infer conclusions concerning the total population? (3) How reliable are these conclusions? (4) How should samples be taken in order that they may be as illuminating and dependable as possible?

Questions 1, 2 and 3 are the basic questions one asks in trying to draw the most useful conclusions, about the population, from the evidence of the sample. Question 4 (which in a temporal sense is out of place and should come first) questions how we should gather our samples. That is, Question 4 leads to all the theoretical development of the *design of experiments.*

Question 1 pretty well covers the subject matter of elementary statistics. Tables, graphs, bar and pie diagrams and the schematic pictorial representations which can be so useful (and sometimes so deceptive[4]) are all ways of summarizing the evidence of a sample. Averages and other related quantities — arithmetical means, medians, modes, geometric means, harmonic means, quartiles, deciles, and so on — are useful for similar purposes; and these also must be used with discrction if they are to be really illuminating. The arithmetical mean income of a certain Princeton class

FIG. 45

five years after graduation, for example, is not a very useful figure if the class happens to include one man who has an income of half a million dollars.

Descriptive statistics of this sort is concerned with broad and vague questions like "What's going on

[4]See *How to Lie with Statistics*, by Darrel Huff and Irving Geis. (New York: W. W. Norton Company, 1954.)

here?"; and the answers returned are a not unworthy example of "doing one's damndest with one's mind, no holds barred," to use Percy W. Bridgman's phrase. It is only when we pass to questions 2,3, and 4, however, that we get to the heart of modern mathematical statistics.

These three questions have to do with different aspects of one common problem: namely, how much can one learn, and how reliably, about a population by taking and analyzing a sample from that population? First of all, what sort of knowledge about a population is possible?

Remember that a population, as one uses the word in statistics, is a collection — usually a large or even infinite collection — of numbers which are measurements of something. It is not possible in the case of an infinite collection, and usually not feasible in other cases, to describe one at a time all the individual measurements that make up the population. So what one does is to lump similar or nearly similar measurements together, describing the population by telling what fraction of all measurements are of this approximate size, what fraction of that size, and so on. This is done by stating in a table, a graph, or a formula just what fraction of the whole population of values falls within any stated interval of values. When this is done graphically, the result is a frequency curve, which describes the distribution of measurements in the population in question. The most widely useful population distribution is the so-called normal or Gaussian probability distribution, which we have met in Chapter XII. A frequency curve can, of course, be described by stating its mathematical equation.

It is frequently useful to give a condensed description of a distribution. If circumstances make it necessary to be content with only two items of information, then one would usually choose the *average* (which the statistician calls the arithmetical mean) and the *variance*. The variance, we recall from Chapter XI, is defined as the average of the squares of the differences between all measurements of the population and the mean of the population. It is a very useful measure of the degree of scatter of the measurements, being relatively small when the distribution clusters closely about the mean and relatively large when the distribution is a widely spread-out one. The square root of the variance, as we know is called the standard deviation. A small variance always means, of course, a small standard deviation, and vice versa.

In the special case of normal distributions, a knowledge of the mean, *mu*, and of the standard deviation, *sigma*, is sufficient to pick out of all possible normal distributions the specific one in question. More complicated distributions may depend upon more than two parameters.

Using the notions just introduced, we can now restate our last three questions: (2) Using the evidence of a sample, what can one say about the population distribution? (3) How can one characterize the reliability of these estimates? (4) How can one select the sample so as to produce the most reliable estimates?

What Sort of Answers Can Statistics Furnish?

Before going on to indicate the answer modern

statistical theory can give to these three questions, it would be well to stop a moment to consider once more, and somewhat more accurately now, the relation between the descriptive statistician who deals only with our original Question 1, and the mathematical statistician who deals with Questions 2, 3, and 4.

In seeking to summarize and describe a sample, the descriptive statistician is in fact trying to shed some light, however dim and indirect, on the nature of the population. Thus he is often trying to give some sort of informal and loose answer to Question 2; and he frequently succeeds in a really useful way. He differs from the mathematical statistician in that he uses only elementary mathematical tools and is therefore unable to give any really precise answers to Question 2, or any answers at all to Questions 3 and 4.

The problem of drawing inferences concerning a population from a sample is a problem in probability. There is an obvious analogy between this procedure and the artificial case of sampling the box of colored balls. It is important to remember that when you take a sample of colored balls out of an unknown mixture, you cannot make simple probability statements about the mixture unless you start out by having some idea about what is in the box. In technical terms, this means that you have to have knowledge, before the drawing of the sample, of the *a priori* probabilities of all possible mixtures that might be in the box. As we have mentioned earlier, Bayes' theorem furnishes a basis for modifying this prior opinion, but it is powerless to originate an opinion.

Bayes' theorem furnishes a sound and simple procedure. But unfortunately it is very seldom applicable to really serious problems of statistical theory, for the good reason that in such situations one seldom has any positive knowledge of the *a priori* probabilities. Consequently, it is necessary for statistics to take recourse to more complicated and more subtle theorems.

It is evident that the statistician can never say for certain what the parent population is merely by sampling it, because the samples will vary. If, for example, you draw from a mixture containing 60 per cent white balls and 40 percent black, you will by no means get this 60–40 ratio of white to black in every sample you take. However, for a given kind of parent population and with suitable methods of sampling, it is possible to work out theoretically the pattern of variation for samples. This knowledge of the pattern of sample variability gives the statistician a toehold. It permits him to look at samples and draw inferences about the parent population.

The pattern of variation depends not only on the population being sampled; it also depends sensitively upon the method of sampling. If you were interested in family sizes in the United States and took your evidence solely from houses with eight rooms or more, obviously the sample pattern would be atypical. When samples are deliberately selected in an atypical way, we call that rigging the evidence. But often samples have quite innocently been taken in an atypical way, as Dr. George Gallup will remember.

It turns out that in general the only good method of sampling is a random method. In a random method the sample is picked in accordance with purely probabilistic criteria, personal choice or prejudice being completely excluded. Suppose television tubes are passing an inspector on a moving belt and it is desired to test on the average one of every six tubes in a random way. The inspector could throw two dice each time a tube passed him and take off the tube for test only when he threw a double number. This, of course, would happen on the average once in six throws of the dice. The tubes thus selected would be a random sample of the whole population of tubes.

The Variation of Random Samples

Now we must note an important fact about the pattern of variation of random samples. Suppose you have a parent population which is normally distributed, with a certain mean value and a certain standard deviation. Suppose you take from this population random samples consisting of a certain number of items, n. Compute the mean for each sample. You will find that this new population of means is normally distributed, just as the parent population was. Because an averaging process has entered in, it is more tightly clustered than was the original population. In fact, its standard deviation is found by dividing the standard deviation of the parent population by the square root of n, the number of items in each sample. Thus if the samples each contain 64 items, the standard deviation of the means of these samples will be one-eighth of the standard deviation of the parent population.

The fact that samples from a normal population have means which are themselves normally distributed tells us that normally distributed populations have a kind of reproductive character. Their offspring (samples) inherit their most important character (normality). And it is comforting to know further that if large samples are taken from almost any kind of population, their means also have almost normal distributions.

The importance of sample pattern can easily be illustrated by a concrete example in manufacturing. A manufacturer makes large numbers of a part which in one dimension should measure one inch with high accuracy. Random samples of the product are measured. If the samples consistently average more than one inch, he knows there is some systematic error in his manufacturing procedure. But if the mean of the sample is just an inch, and the variations from the mean fall into the pattern of distribution which is theoretically to be expected when the parent population itself is normally distributed about an average of one inch, the manufacturer can conclude that systematic errors have been eliminated, and that his manufacturing process is "under control."

Sampling is not merely convenient; it is often the only possible way to deal with a problem. In the social sciences particularly it opens up fields of inquiry which would otherwise be quite inaccessible. The British Ministry of Labor was able to carry out a most useful study of working-family budgets for the entire nation from detailed figures on the expenditures of only 9000 families over a period of four weeks. Without sampling such studies would be wholly impracticable.

Questions (2) and (3): Statistical Inference

To return to our main argument. We are now

better prepared to deal with Questions (2) and (3).

Suppose that the mean lifetime of a certain type of electronic tube is known to be 10,000 hours and the standard deviation 800 hours. The engineers now develop a new design of tube. A sample of 64 of the new-type tubes is tested, and the mean life of the 64 tubes in the sample is found to be 10,200 hours — 200 hours longer than the mean life of the old population.

Now the new design may actually be no longer lived than the old. In that case the sample of 64 just happened to be a somewhat better than average sample. Clearly what the engineers want to know is whether the apparent improvement of 200 hours is real or merely due to a chance variation.

The amount of variation one expects from chance can be estimated by comparing the actual deviation with the standard deviation. Since the standard deviation of the means of samples of 64 items is one-eighth the standard deviation of the parent population, in this case the standard deviation in the means of such samples would be one-eighth of 800, or 100 hours. Hence the apparent improvement of 200 hours in our sample of new-type tubes is twice the standard deviation in the means of such samples.

Probability theory tells the statistician that the odds are 19 to 1 against a difference of this size between the sample and population means occurring merely by chance. He therefore reports: "It seems sensible to conclude that this mildly rare event has not occurred, and that on the contrary the sample of 64 in fact came from a new population with a higher mean life. In other words, I conclude that the new design is probably an improvement."

This is one of the common ways of dealing with

such a situation. There are various rather complicated and subtle weaknesses in the argument just given, but we need not go into them here. A more satisfactory way of dealing with the same problem would be to apply the modern theory of statistical estimation, which involves the use of so-called confidence intervals and confidence coefficients. Here the statistician proceeds as follows: He says that the sample of 64 tubes comes from a new population which, while assumed normal, has an unknown mean and variance; and he very much wants to know something about the mean of this population, for that information will help him to conclude whether the new design is an improvement.

Now we must remember that statistics deals with *uncertain* inference. We must not expect the statistician to come to an absolutely firm conclusion. We must expect him always to give a two-part answer to our question. One part of this reply goes: "My best estimate is . . ." The inescapable other part of his reply is: "The degree of confidence which you are justified in placing in my estimate is . . ."

Thus we are not surprised that the statistician starts out by choosing a number which he calls a confidence coefficient. He might, for example, choose the confidence coefficient .95. This means that he is about to adopt a course of action which will be right 95 per cent of the time on the average. We therefore know how much confidence we are justified in placing in his results. Having decided on this figure, statistical theory now furnishes him with the width of a so-called confidence interval whose midpoint is the mean of the sample. In our example this interval turns out to be 10,200 plus or minus 195 hours, or from 10,005 to 10,395 hours. The statistician then answers Questions (2) and (3) as follows: "I estimate that the mean of the population of lifetimes of new-design tubes is greater than 10,005 hours and less than 10,395 hours. I can't guarantee that I am correct; but in a long series of such statements I will be right 95 per cent of the time. Since this range is above the mean life of the old tubes, I conclude that the new design is probably an improvement."

If the statistician had originally decided to adopt a procedure that would be correct 99 per cent of the time, his confidence interval would have turned out wider. He could make a less precise statement but make it with greater confidence. Conversely, he could arrange to make a more precise statement with somewhat less confidence.

Finally, let us examine this same question in the still more sophisticated manner that goes under the name "testing of statistical hypotheses." Here one starts by making some sort of guess about the situation, and then goes through a statistical argument to find out whether it is sensible, and how sensible, to discard this guess or retain it.

Thus the statistician might tentatively assume that the new design is equivalent to the old in average tube life. Of course, he hopes that this is not true. Although it sounds a little perverse, it is in fact customary to start with a hypothesis that one hopes to disprove.

Here, just as in the previous case, the statistician first picks out a number which is going to tell what confidence we dare have in his statements. Actually he uses something which might be called an "unconfidence coefficient," for it measures the per cent of the time he expects to be wrong, rather than the per cent he expects to be right.

This unconfidence coefficient is technically called the significance level. Let us say that the statistician chooses a significance level of .05, which is exactly equivalent to .95 as a confidence coefficient. Then he calculates the confidence interval for this confidence coefficient. Since we have the same confidence coefficient as before, we already know that this particular confidence interval reaches from 10,005 hours to 10,395 hours. The statistician then reports: "The mean life of the old population (10,000 hours) does not fall within my confidence interval. Therefore theory tells me to discard the assumed hypothesis that the new tube has the same average lifetime as the old. The assumed hypothesis may of course actually be true. But theory further tells me that in cases in which the hypothesis is true, and in which I proceed as I just have done, I will turn out to make mistakes only 5 per cent of the time."

This report, if one thinks it over carefully, is rather incomplete. It says something about the probability of one sort of error — the error of discarding the hypothesis when it is in fact true. But it says nothing about another sort of error — accepting the hypothesis when it is in fact false. In certain situations one of these two mistakes might be very dangerous and costly and the other relatively innocuous. There are available still more refined statistical procedures (called the Neyman-Pearson methods and the theory of decision functions) in which one designs the test so as to make a desirable compromise with respect to the probabilities of the two types of error.

Question (4): Experimental Design

So far we have given no sort of reply to Question (4) — How should samples be taken in order that they may be as illuminating and dependable as possible?

It will not be feasible, within the limits of this small book, to give more than a hint of the answer to this question.

The use of statistical theory to plan the design of experiments began about 1919 when, the then young mathematician R. A. Fisher (later, Sir Ronald) went to the great Rothamstead Experimental Station in England, and started work which led to his now classic *Statistical Methods for Research Workers*, which first appeared in 1925.

The central purpose of the theories of experimental design which were founded by Fisher, and which have led to a vast and vastly useful body of knowledge,[5] was to maximize, for a given experimental effort, the amount and reliability of the information which can be obtained.

When one wishes to test, for example, one or more fertilizers for their capacities to increase the yield of

FIG. 46

[5]See, for example, *Experimental Designs*, by W. G. Cochran and Gertrude M. Cox (New York: John Wiley & Sons, 1950).

some crop plant, it is necessary to use different plants, growing on different spots of ground. However carefully the test plots are chosen and prepared, there will be some variation in results due to numerous factors other than the fertilizer. This is obvious: for if you use the *same* fertilizer, in the same doses and on the same schedule, on several small test plots the yields of the individual plots will of course not be the same.

The modern technique of statistical experimental design recognizes and faces this matter of the inevitable variation due to influences *not* under study as well as those due to the influences which *are* under study. The essential trick is to arrange the experiment in such a way that when the data are finally in hand for analysis it is possible by the statistical methods to sort out that part of the variation which is due to the influence or influences under study (which might, for example, be type of fertilizer, dose used, and times of application) from that part of the variation which is due to uncontrollable influences not under study (which might include minor differences in soil, drainage, microclimate, subsurface conditions, etc.).

This sounds like quite a trick, doesn't it? In fact it *is* quite a trick. But it is of the highest usefulness, especially in many fields of agricultural, biological, and medical experimentation, where one is condemned to the use of test objects which cannot be really identical. And it is also of great use in many industrial fields. Lady Luck does not spend all her time, by any means, frittering away the hours in careless or reckless play. She is a dependable servant of all experimental science.

Chapter XV

PROBABILITY AND GAMBLING

Let the King prohibit gambling and betting in his kingdom, for these are the vices that destroy the kingdoms of princes.

THE CODE OF MANU (about A.D. 100)

It cannot be denied that Lady Luck was born in the gaming rooms of the seventeenth century. She presumably had her crib lined with green felt, and used a pair of dice for a rattle. The early history of the mathematical theory of probability was to a large extent, as we have seen in the opening chapters of this book, a record of the solving of one betting problem after another.

There is a great deal more to gambling than the mathematical probabilities involved. It can be amusing, exciting, rewarding, or disastrous. It can lead to a champagne dinner, a pearl necklace, or suicide. It can be the very occasional fun of a light moment, held within the bounds of reasonable costs for entertainment; or it can become a compelling addiction, wasting money essential for serious purposes.

The results of probability theory have undoubtedly been used, and effectively used, to aid clever and knowledgeable gamblers. If you read *The Education of a Poker Player*, by the very able Herbert O. Yardley, who was a famous cryptographer,[1] you will certainly conclude that a person who understands the odds which obtain in various circumstances has a large and profitable advantage over the person who does not. A number of bridge experts have become adept at the application of probability to their game.[2] But strategy, bluffing, the technique of legitimate communication between partners, intuition, audacity — all these play very important roles.

The Game of Craps

The game of craps furnishes a good example of probability calculations in a gambling game; for craps is sufficiently more complicated than "heads and tails" to raise some nice little problems, but not so complicated (as is bridge, for example) that the calculations are tedious. Craps, also, is a pure game of chance, involving no element of skill,[3] and involving judgment only in connection with decisions concerning entering and leaving the game.

If you are not familiar with the game, look back to page 281 for a brief description. After the first roll the game is either terminated in an immediate win, or terminated in an immediate loss, or continued. The

[1]He was largely responsible for breaking the Japanese diplomatic code in World War II. His book was published by Simon and Schuster, 1957.

[2]See, for example, Oswald Jacoby's *How to Figure the Odds* (New York: Doubleday & Company, 1947), and *Théorie Mathématique du Bridge*, by F. E. J. Bard and A. Chéron (Paris: Gautier-Villars, 1940).

[3]There will, of course, be protests against this remark!

FIG. 47

probabilities of these three cases are seen to be, respectively,

$$6/36 + 2/36 = 8/36,$$

$$1/36 + 2/36 + 1/36 = 4/36,$$

$$3/36 + 4/36 + 5/36 + 5/36 + 4/36 + 3/36 = 24/36$$

If the first roll produces a 4, 5, 6, 8, 9, or 10, so that the game continues, then the player subsequently wins if he "makes his point" (i.e., repeats his first roll) before he "craps out" by throwing a 7. If the 7 turns up before he duplicates his first roll, he loses.

Having established his point by his first roll, from then on — until this portion of the game is completed — the player has but two critical numbers, his point, and 7.

What is the chance that he will win by making his point before he rolls a 7? Let's turn aside for a moment and solve a little problem in probability theory.

Suppose, for example, that you have series of trials on each of which *three* mutually exclusive things can happen — result A with probability a, result B with probability b, and result C with probability $c = 1 - a - b$. (We are thinking, of course, of A meaning "I make my point"; of B meaning "I roll a 7 and lose"; and C meaning "I roll neither my point nor 7, so I go on to roll again.")

What is the probability that, if I keep on with these trials the outcome A occurs before the outcome B?

This can occur in any one of the following mutually exclusive ways —

On trial #1 by A occurring then: for which the probability $= a$

On trial #2 by C occurring on trial #1 and A on trial #2: probability $= c \cdot a$

On trial #3 by C occurring on trials #1 and #2, and A on trial #3: probability $= c^2 \cdot a$

etc.

Thus the total probability that A occurs before B is

$$a + ac + ac^2 + ac^3 + \cdots = a(1 + c + c^2 + c^3 + \cdots)$$

which[4] is

$$\frac{a}{1 - c} = \frac{a}{a + b}$$

There is another way to see that this value is correct. No matter how many times C occurs (i.e., a number is rolled which is neither the point or 7), on

[4]Look back at the footnote on page 168.

the next roll the *odds* for A (the point) as compared to B (crap with 7) are as a is to b. These being the relative odds on each successive opportunity for the matter to be settled, these are also the over-all odds for success (that is, for A occurring before B). If the relative odds favoring A over B are as a to b, then the probability of A occurring before B is $a/(a + b)$, as before. It isn't necessary to worry over the contingency that neither A nor B will occur — that neutral numbers (case C) will come up indefinitely. For the probability c of C is less than 1, and the probability c^n, that C keeps on for n successive times, approaches zero as n keeps on getting larger and larger.

We are now in a position to calculate the probability that a player will make his point before being crapped out. For the points 4 or 10, the probability of making either is, as we have seen, 3/36, while the probability of throwing 7 is 6/36. Thus the probability of throwing a 4 before throwing a 7 is

$$\frac{3/36}{3/36 + 6/36} = \frac{3}{9} = 0.333$$

Similarly the probability of throwing a 5 or 9 before a 7 is $4/10 = 0.400$; and the probability of throwing a 6 or 8 before a 7 is $5/11 = 0.454$. The 4 and 10 are hard points to make ($p = 0.333$) as compared with the medium-hard points 5 or 9 ($p = 0.400$), and as compared with the relatively easy points 6 or 8 ($p = 0.454$).

With these figures available, it is now easy to calculate the probability, as he picks up the dice to make his initial roll, that the player will win. He can win in either of two mutually exclusive ways — by winning on the first roll, or by establishing his point on the first roll and then subsequently making that point. The first of these ways has a probability, as we saw above, equal to 8/36.

To this we have to add the probability of making his point on a roll subsequent to the first roll. Thus the total probability that the player will win is

$$\frac{8}{36} + 2\cdot\frac{3}{9}\cdot\frac{3}{36} + 2\cdot\frac{4}{10}\cdot\frac{4}{36} + 2\cdot\frac{5}{11}\cdot\frac{5}{36} = \frac{244}{495} = 0.4929$$

the second term being due to the two possible initial rolls 4 and 10, the third term to the two possible initial rolls 5 and 9, and the fourth term to the two possible initial rolls 6 and 8.

This is a very interesting result. Most crap players like to be the one doing the rolling, presumably be-

FIG. 48

cause they "feel lucky," think the dice are "hot," and enjoy the active role of the player. But their chances are, nevertheless, not quite equal to the 1/2 value which characterizes a fair game. In the long run the crap shooter will lose about 1.41 per cent of the amounts staked, this being slightly worse for the player than the usual "house advantage" (1.35 per cent) in playing even or odd numbers in roulette at Monte Carlo (where the wheels have one zero), almost twice as good for the player as the usual house advantage (2.7 per cent) when playing individual numbers at roulette, and roughly four times as good for the player as the house advantage when playing even or odd numbers at roulette if the wheel has two zeroes.

The Ruin of the Player

Although gamblers have undoubtedly learned a great deal from probability theory, they clearly never learn — or in any event never accept — the main lesson probability theory can teach about gambling. For that lesson is, in brief, "If you keep on gambling, you will lose."

One of the most simple, and yet most impressive, illustrations of the results of continued gambling is furnished by an old and classical problem. It was studied as long ago as the 17th century by James Bernoulli; and in various forms of increasing generality it has been studied right up to the present day.

Consider a player, whom we will call P, who carries on a series of bets with a gambling house H. Each bet will be for one dollar, and we will suppose a *fair* game in which, on each trial, the gambler has an even chance of winning. It will simplify things for us, and not affect the problem in any significant way, to suppose that the total amount of money at stake in this game is a fixed amount of say C dollars (C for "capital"). Of this total capital the player P starts the game with say d dollars, whereas the house H starts out with say D dollars, with $d + D = C$. The small letter d and the large letter D indicate that the player presumably has smaller resources than those of the house.

We will suppose that this game proceeds until one of the two, player or house, is ruined. That is, the game keeps on until the player's capital hits zero, or has increased to C, so that he has "broken the bank at Monte Carlo." We want to calculate the probability that the player will be ruined, and also the probability that he will break the bank.

We will denote by R (R for "Ruin") the probability that the player ultimately will be ruined; and when we want to call attention to the fact that his chance of ruin depends on the size of the capital he starts with, we will write this with a subscript, as R_d. Then R_{d+1} would, for example, be the probability of ruin when he has a capital one dollar larger than d (and the house one dollar less than D, since we consider C fixed).

We will solve our problem by using a little trick which is often useful in probability calculations, and which may be, for you, a new and therefore interesting procedure.

We know that before he makes his first trial the player's probability of ruin is R_d. Let us, at that same moment before the player has made any bet, calculate this same probability R_d in a different manner, thus obtaining an expression which must be equal to R_d.

We do this by thinking of the two ways in which the first trial might eventuate. There is a probability of 1/2 that after this trial the player will have capital $d + 1$ and therefore then have a probability of ruin equal to R_{d+1}. But there is also a probability of 1/2 that his capital will be reduced to $d - 1$ so that he then will have a probability of ruin equal to R_{d-1}. If we write out our present estimate of the probability of ruin, but express this in terms of the possible results of the first trial, we would say that the expected value of the ruin probability is

$$\frac{1}{2} R_{d+1} + \frac{1}{2} R_{d-1}$$

At this same moment, before any trial, we can calculate the expected value of R_d in another way; that is, we are certain (probability equals 1) that the player's capital is then d, so the expected value of R_d as estimated at that moment can also be written

$$1 \cdot R_d = R_d$$

These alternative, and both correct, expressions for the expected value of R_d must be equal, so

$$R_d = \frac{1}{2}(R_{d+1} + R_{d-1})$$

You may very well never have met this kind of equation. It is called a "difference equation" because it relates the values R must have for different values of the parameter which occur as a subscript. This equation defines the way in which R_d must depend on d. But the parameter d can take on only *integral* values, so this equation is not at all like the ordinary linear equations in algebra, which involve an unknown x which can take on *any* value.

Our difference equation cannot be explicitly solved unless we add a little more information — what in mathematics are called[5] "boundary conditions." Namely, we know that when $d = 0$ the gambler is certainly ruined, so that

$$R_0 = 1$$

And we also know that if and when his capital is increased to C he has broken the bank and his chance of being ruined has vanished. Thus

$$R_C = 0$$

We will now solve the difference equation by what mathematicians solemnly call "the method of particular solutions," but which really means "by guessing the general form of the answer." Thus we guess that our equation and boundary conditions can be satisfied by a simple linear expression

$$R_d = A + B \cdot d$$

where A and B are constants (that is, quantities which do not depend on d); and where the two

[5]An hour or so after I wrote this sentence, my wife and I were driving northward from Florida on Route 301. She was looking at a folder which listed, in a vertical column, a lot of towns or cities on this route. The adjoining column was headed "Distance between cities" — but each distance was *directly opposite* a city. Was this the distance from the city in question *from the preceding one*, or *to the succeeding one?* This question could be settled at once by looking at either the first or the last entry, which thus served as a sort of "boundary condition."

constants A and B are to be determined by the conditions that

$$R_0 = 1 \qquad \text{or} \qquad 1 = A + B\cdot 0$$

$$R_C = 0 \qquad \text{or} \qquad 0 = A + B\cdot C$$

These two easy equations (and now this *is* ordinary algebra give at once

$$A = 1$$

$$B = -\frac{1}{C}$$

and our solution is

$$R_d = 1 - \frac{d}{C} = \frac{D}{d + D} \tag{34}$$

Now, by merely starting out with different agreements as to the notation to be used, we would have arrived, for the probability that the *house* will be ruined, at the expression

$$1 - \frac{D}{C} = \frac{d}{d + D}$$

The sum of the last two probabilities is

$$\frac{d + D}{d + D} = 1$$

Hence it is *certain* either that the player will be ruined or that he will break the bank so that the house is ruined. This sort of game in other words, always results in complete victory for one side or the other, and never in a tie.

The expression for the probability that the player breaks the bank, namely $d/(d + D)$, and for the probability that the house breaks the player, namely

$D/d + D$, are very natural and reasonable. The chances (to be ruined) of the player are to the similar chances for the house as d is to D. That is to say, their relative chances to succeed ultimately are proportional to their resources.

Next, take a sharp look at equation (34). If the house, which of course may be backed financially by a very wealthy syndicate, has very large resources D as compared with the player's resources d, then

$$\frac{d}{C} = \frac{d}{d + D}$$

is a very small number, and R_d is accordingly very near unity. Hence the solemn conclusion: If you play a game against a house (professional gambler or syndicate) which has resources which are large as compared with your own, then you are exceedingly likely to face ultimate ruin, *even though the game is itself a fair one*, with no advantage to the house on individual plays.

On the other hand, let us calculate the expected gain of the player who undertakes carrying this sort of contest to its conclusion. His ultimate loss may be $-d$, the probability of this loss of his whole pile being the probability of ruin R_d. Or he may break the bank (the probability of this being $1 - R_d$) and gain the house's whole stake D. Thus the player's expected ultimate gain is

$$- d \cdot R_d + D(1 - R_d) = C(1 - R_d) - d$$

If we set this equal to zero, we get

$$R_d = 1 - \frac{d}{C}$$

which is precisely our previous equation (34), which holds only if $p = q = 1/2$. That is, the game which is "fair" for each individual trial remains "fair" (i.e., with zero expected gain) when carried to conclusion.

It may seem paradoxical to you that the game is, on the one hand, fair, whereas, on the other hand, the probability of the player's ruin should be so large if his adversary's capital is very large as compared to his. As the adversary's capital is made larger and larger, the chance that the player will be ruined is increased and the probability that he will break the bank is decreased. But if he *does* break the bank, his win, relative to his own capital, is increased, and all these factors just balance out, making the total game "fair."

If we consider the same situation of continuing wagers between player and house, the probability of the player's winning a dollar on any one round now being p, and of his losing a dollar being $q = 1 - p$, but it not now being assumed that p and q have the fair values 1/2, then the difference equation is somewhat altered and the solution is a more complicated one, although the procedure for solving this more general case is exactly similar to the procedure we have just gone through. In fact, it turns out that R_d, the probability of ultimate ruin of the player, is now given by

$$R_d = \frac{\left(\frac{q}{p}\right)^C - \left(\frac{q}{p}\right)^d}{\left(\frac{q}{p}\right)^C - 1} \qquad (35)$$

Again it checks out that the probability of the player's being ruined plus the probability of his breaking the bank add up to exactly 1 or certainty; so again it is true that the game always ultimately proceeds to a complete victory for one or the other, with no ties.

From equation (34) it can be calculated that for a game fair on each trial ($p = q = 1/2$), the probability R_d of the player's ruin is

The Player's Resources d	*The Resources D of the House*	*The Probability R_d that the Player Be Ultimately Ruined*
9	1	0.1
1	9	0.9
1	99	0.99
1	999	0.999

For all these cases, the over-all expectation of the player is zero. In the last case, for example, he has 999 chances out of 1000 of losing his one dollar, but one chance out of 1000 that he will win the $999 which the house has, and these two expectations just balance out to zero.

On the other hand, suppose the individual trials are more or less unfair to the player (p less than 1/2), as would be the case in any practical case of gambling against a professional. Then we must use the more complicated equation (35).

Now if the game is unfavorable to the player, so that p is less than 1/2, the ratio (q/p) will be greater than unity. If the total capital in the game — namely C units of money, using the basic bet per trial as the unit — is fairly large, then the quantity $(q/p)^C$ is large enough so that it really doesn't matter, in the denominator of (35), whether or not we subtract the 1. Neglecting this 1, the formula simplifies a lot, for

$$\frac{\left(\frac{q}{p}\right)^C - \left(\frac{q}{p}\right)^d}{\left(\frac{q}{p}\right)^C} = 1 - \left(\frac{q}{p}\right)^{d-C}$$

$$= 1 - \frac{1}{\left(\frac{q}{p}\right)^D} \qquad (36)$$

That is to say, the probability that the player will be ruined differs from 1 (certainty) by a quantity which is awfully small if the resources D of the house are large. The player's probability of ruin comes closer and closer to certainty as the resources D of the house are increased.

Let's look at three cases, starting in Case I with a situation which is only very mildly adverse to the player, taking as Case II a situation in which his odds are 47 to 53 (still not a very adverse game), and taking as Case III a situation in which the player's odds are only 40 to 60. We will see in a minute that these three cases pretty well cover the range of actual games the player might wish to play. And we will also consider, for each of these three types of games, three cases as far as the resources of the player are concerned. First we will take $d = D = 5$, so $C = 10$, as a case of a player who is gambling with a friend whose resources are equal to his (although the "friend" has an edge in his favor as regards the probability). Secondly we will take $d = 10$ and $D = 40$, or $C = 50$, as the case of a player indulging in an unfavorable game with a professional who has resources four times as large as his. Finally we will take $d = 10$ and $D = 90$, or $C = 100$, as a case of a real sucker playing an unfavorable game with a house whose resources are nine times his.

Here is a little table:

	Case I	*Case II*	*Case III*
Odds for the player	49	47	40
Odds for the house	51	53	60
Probability of player's winning individual trial	.49	.47	.40
Value of (q/p)	1.041	1.128	1.5
Value of $(q/p)^C$ when $C = 10$	1.49	3.32	57.7
Value of $(q/p)^C$ when $C = 50$	7.39	407	6.4×10^8
Value of $(q/p)^C$ when $C = 100$	54.6	1.7×10^5	4.1×10^{17}
Value of $(q/p)^d$ when $d = 5$	1.22	1.826	7.59
Value of $(q/p)^d$ when $d = 10$	1.49	3.32	57.7

We can see at once that for $C = 100$ the neglect of the 1 in the denominator is completely justified for Cases II and III (for 1 is negligibly small as compared with 1.7×10^5 or 4.1×10^{17}) and causes no serious error (1 is about 1.8 per cent of 54.6) for Case I. For $C = 50$ we can use the approximate formula (36) for Cases II and III: and when $C = 10$ we can use it only for Case III.

These figures enable us to proceed to the table we are really interested in (Table XIII):

Roulette, Lotteries, Bingo, and the Like—

As to the probabilities in various games you might consider playing, we have already seen that for craps the person who is rolling has a probability of winning which is 0.4929.

In roulette it makes a difference whether you play at Monte Carlo, where there is one zero on the wheel, or in the United States, where there are usually two zeroes. The odds at roulette depend, of course, on the bet you make. *Red, black, even, odd, "manque"* (all numbers up to 18), and "*passe*" (all numbers over 18) are even-money bets in that you win (if you win) what you bet. But if zero comes up, the bet goes "in prison" and goes to the house if you lose the next round, or is returned to you if you win. The over-all probabilities in roulette, however, are around 0.49, a little more for the even-money bets on a one-zero wheel, a little less for bets on individual numbers, and of course less if there are two zeroes.

When you are concerned with a very complicated game it is often exceedingly tedious to try to figure out all the probabilities of all the different aspects of

TABLE XIII

PROBABILITY OF THE GAMBLER'S RUIN

Resources d of the player, and D of the house	*Probability of Player's Ultimate Ruin*		
	Case I, $p = .49$	*Case II*, $p = .47$	*Case III*, $p = .40$
$d = 5$, $D = 5$	0. 55	0. 65	0. 88
$d = 10$, $D = 40$	0. 92	0. 99	$1 - 9 \times 10^{-8}$
$d = 10$, $D = 90$	0.99	$1 - 2 \times 10^{-5}$	$1 - 10^{-16}$

the game. In such cases it is useful to go right to the over-all figures of "take."

If in a game you bet \$1 each time, with a probability of winning an extra dollar equal to say m/N, then in N bets you will (on the average, of course) win m times, giving you a sum of $2m$ dollars (the return of your bets plus the dollars you won), and you will lose $N - m$ times. You end up a sequence of N bets with $\$2m$ and the house with a take of $\$(N - 2m)$. Thus in terms of take

$$\frac{\text{take}}{\text{total amount bet}} = \frac{N - 2m}{N} = 1 - 2\frac{m}{N} = 1 - 2p$$

and from this

$$p = \frac{1}{2}\left[1 - \frac{\text{take}}{\text{total amount bet}}\right]$$

$$= \frac{1}{2}\left[\frac{\text{what you have left}}{\text{total amount bet}}\right]$$

That is, if the customers invest \$10,000 and return home with \$8000, so that the take of the house is \$2000, then this is equivalent to a simple game in which every time you bet a dollar you have a probability of winning equal to $1/2(.8) = 0.4$. In either case the expectation of the gambler is the same.

For a usual type of slot machine we have seen, on page 159, that for \$8000 invested the players take home \$5888. The over-all probability is thus about 0.37. For horse racing the take on individual races usually varies from 12 to 18 per cent, leading to probabilities ranging from 0.41 to 0.44. But because winners very often re-bet their gain, it turns out that a track on a certain day took in \$580,700, and paid back \$240,700 so that the over-all probability of an optimistic and determined better was only about 0.20.

Some government lotteries are based on rather moderate takes of say 10 per cent, leading to $p = .45$; but many others are, of course, less favorable. The Irish Sweepstakes is supposed to operate on a take of about 35 per cent. Facts about the Mexican Government lottery are, understandably, not readily available; but an experienced person who has attended many of their drawings estimates that their take is at least 50 per cent.

Football pools in this country have shockingly large takes. The Swedish Government football pool is, I believe, run on the basis of a 50 per cent take.

Bingo at Reno and Las Vegas is said to involve only a 20 per cent take, but the bingo take under other auspices varies from 25 per cent up to almost any figure you care to name. Professor Fox of the University of Wisconsin, who is an ardent student of such matters, has written me "Bingo can produce anything, and I always say that God only knows, and that He ought to know because most of it is played in church basements."

Table XIII thus pretty well spans the games you might think of playing. Its advice should be clear: Play only if you can afford to charge the loss up to entertainment.

In the "friendly" game in which the resources are equal, the player with the adverse odds is always more likely to be ruined than not, and if his probability is as low as $p = .40$, the odds are better than seven to one that he will be ruined. When the player takes on the professional with four times his resources, his probability of being ruined goes up mildly for Case I with its game which is *almost* fair;

but goes up disastrously for Cases II and III. As for playing against the house with nine times his resources (really not a very wealthy house), the player obviously should have stood in bed. His chances of ultimate survival, in the three cases, are one in a hundred, one in fifty thousand, and one in ten thousand million million!

From the formulas given above, and from similarly derived formulas for the probable length of game that will occur before the player is ruined or the bank broken, various interesting results follow:

For instance:

The entire series of trials constitutes an over-all fair game if, but only if, the individual trials are themselves fair.

Reducing the stakes — i.e., reducing the amount bet on each trial — has just the same effect on the probability of ruin as keeping the stakes the same but increasing the capital in the inverse ratio. That is, reducing the bet per trial from $1 to 25 cents has the same effect as leaving the bet per trial unchanged but quadrupling the capital.

On the other hand, if the stakes are doubled and the initial capital amounts d and D are unchanged, the probability of ultimate ruin decreases (are you not surprised at this!) if the player's probability on each trial is less than 1/2.

The expected number of trials in a fair game which is carried to conclusion (ruin of player or house) is dD, the product of the number of units of money put into the contest by the player and house. Thus if a player with $100 plays heads or tails with $1 bets against a house which has $1000, carrying the contest to conclusion, the expected number of tosses is 100 · 1000, or 100,000.

In a game disadvantageous to the player, so that his probability of success per trial p is less than his probability of failure q, the expected number of trials in a contest carried to conclusion is always less than $d/(q - p)$. Thus with the same initial capitals as in the last illustration (d = \$100, D = \$1000) but with $p = 0.4$ and $q = 0.6$, the expected duration is less than $100/0.2 = 500$ trials. If the house has a capital D which is large as compared with the player's capital d, then the expression $d/(q - p)$ gives a good approximation to the expected duration of the game. It depends only on the player's capital (and the probabilities) and not on the capital of the house. This is all right, for if D is large compared with d and if the game is unfavorable to the player, then it is *highly* probable that the player will be wiped out, and the size of the resources of the house does not come under consideration, for the house is never in any real danger of running out of funds.

Gambling Systems

Levinson's book, *The Science of Chance*, to which I have referred several times, contains a rather detailed and very interesting analysis of a variety of gambling systems, including the "Martingale," in which you double your preceding bet every time you lose, and then return to your normal starting bet right after any win.

Suppose you use this doubling system in betting that a tossed coin will turn up heads, starting each time with a one dollar bet.

If heads comes up the first toss, you win a dollar. If tails comes up the first time and heads the second, you lose one dollar on the first toss, but win two on the second, for a net gain of one dollar. If tails comes up twice in a row, and then heads appears, you lose \$1 + \$2, but you win \$4 on the third toss for a net gain of \$1. No matter how long tails persists, when a head eventually appears you net \$1. Your total win is \$1 times the number of times heads appears.

So far this sounds fine. But remember the two aspects of the law of large numbers! The *ratio* of heads to tails tends to approach 1/2; but the *actual excess* of one side over the other tends to grow. Thus the excess of heads over tails may (if you are lucky enough) be large at the moment the game stops, and you win a number of dollars which is substantially more than half the number of tosses.

But (and this deserves a separate paragraph) it is just exactly as likely that the excess of tails over heads may be large. And what does this do to you?

Unless you have large resources and the nerve to commit them in the face of "bad luck," it wipes you out — ruins you. For if you double the bet every time, you run rapidly into big numbers.

It is easy to estimate what sort of adverse runs may occur, for a run of say r uninterrupted tails, preceded by a head and followed by a head so that it is in fact a run of exactly r, involves the specification of how $r + 2$ tosses come out. The probability that $r + 2$ tosses come out in exactly a specified way is, of course, $1/2^{r+2}$. Therefore the expected number of "r-runs" of one side of the coin in n tosses is

$$\frac{1}{2^{r+2}} \cdot n$$

If we consider exactly 1024 tosses (taking this special number because 1024 is 2^{10} so that the multiplications come out evenly) we have this table:

Length of run of tails	1	2	3	4	5	6	7	8
Expected number in 1024 tosses	128	64	32	16	8	4	2	1

These tabulated runs of tails require a total of

$$1\cdot 128 + 2\cdot 64 + 3\cdot 32 + 4\cdot 16 + 5\cdot 8 + 6\cdot 4 + 7\cdot 2 + 8\cdot 1 = 502$$

tosses; and the same number of tosses would, on the average, be used in getting similar runs of heads. So we have, so far, accounted for $502 + 502 = 1004$ tosses. But we are tossing 1024 times! What about the remaining 20 tosses?

The remaining 20 tosses will, on the average, be used up in runs longer than 8, which are less likely but which, of course, may occur.

A run of 8 tails requires you to bet, and lose, the following sums in dollars if you follow the doubling system

$$1 + 2 + 4 + 8 + 16 + 32 + 64 + 128 = 255,$$

and longer runs of tails would require more resources and more nerve.

You can play around with calculations such as these for hours and hours. It is much cheaper than gambling, not as exciting, but I think just as much fun.

And the net result of calculations of this sort — and of the almost countless alternative calculations for other systems — is this: It is easy to devise a system that gives you a high probability of making a small (or even modest) profit over the short run; but the tactics necessary for assuring a good probability for a reasonable profit (which tactics normally involve having a very large capital, playing against a house which has no limit on bets, and keeping your nerve) are precisely the tactics that *also* increase the likelihood that you have, at some juncture, a big loss. You can weather these bad storms by increasing your capital — but if you reach a limit beyond which you cannot or will not go, then you are ruined.

The chance of a small gain can be made large, and the probability of a large loss can thus be made small. But the two risks keep on balancing each other; and if your resources have an upper limit, the probability that you will be eventually ruined when playing against a house prepared for any eventuality approaches certainty as you recklessly and desperately continue. This is just what we saw earlier, in discussing equations (34) and (35).

Mathematicians have used great ingenuity in studying and analyzing this matter of gambling systems. Suppose (still talking about tossing a coin) the player devises *any rule he pleases* to tell him at what juncture he will bet and when he will not bet. Just agreeing that his rule depends only on the result of tosses already made (for he cannot foresee the future, and he can't change bets once they are made), and that his system requires him to *keep on*, then it can be shown[6] that this system is completely futile. It doesn't really affect the player's chances one bit. He might just as well abandon his fancy system, bet every time — and proceed to be ruined.

Gambling systems, says Lady Luck out of her three centuries of practical and theoretical experience, are no good.

[6]Page 187, Will Feller's book, often previously referred to.

Chapter XVI

LADY LUCK BECOMES A LADY

The conception of chance enters into the very first steps of scientific activity in virtue of the fact that no observation is absolutely correct. I think chance is a more fundamental conception than causality; for whether in a concrete case, a cause-effect relation holds or not can only be judged by applying the laws of chance to the observation.

MAX BORN

An embarrassing proportion of key decisions in the Government, from the negotiation of treaties to the management of resources, are made on the basis of insufficient information and unproved assumptions. And this is not so different from the way we conduct our private affairs.

FROM AN EDITORIAL IN *Science*

Preliminary

There is no denying that Lady Luck was born of rich but dishonest parents. There is no denying that her early days were spent with unsavory companions, and that she was then almost wholly concerned with superficial, not to say improper, activities.

But all that is now far away and long ago. To be sure, Lady Luck has never lost her interest in sporting events; but she now appears at Epsom Downs in a lovely French frock, accompanied by the leading figures of society. Among her most enthusiastic modern admirers are now found not only the great sportsmen, but also the leading businessmen, the statesmen, the military experts, the scientists, and the philosophers. Even the theologians drop in occasionally at her teas.

In this last chapter I want to give at least a hint of the role of probabilistic thinking at the present time, to indicate how widespread are its applications and how completely fundamental its influence.

Before citing Lady Luck's qualifications for the intellectual peerage, I want to make a few closing comments concerning the concept of the probability of an event. Many books on probability would — and logically quite properly — begin with a discussion of that sort. It is, however, somewhat easier to discuss such issues after achieving some preliminary acquaintance, even if on our informal basis and without a proper introduction.

Also I want to tuck into this final chapter a few odds and ends which have not fit naturally into the previous chapters but which I can't bear to leave out.

The Probability of an Event

In ordinary speech we have a large number of words and phrases that attempt to express our opinion of our degrees of confidence. We say *probable, likely, plausible, credible, impossible, presumptive, improbable,* etc. We say *have a good chance, there is a reasonable prospect,* etc., etc. The theory of probability seeks to refine these vague notions, and to demonstrate how, from one set of statements which are believable, a person can obtain a lot of otherwise unsuspected statements which are just as valid as the starting statements, but much more interesting and useful.

As we now think back to the way in which we met the concept of mathematical probability, in Chapter III, we can agree that it was naïve of the early founders of the theory to define the probability of an event in terms of "equally probable events." You can say that two events are "equally probable" — that is, that their "probabilities are equal" — only providing you already know what "probability" is and how you measure it. And if you know that, then clearly you are already far beyond the point of defining probability.

In other words, the classical definition in terms of equally probable events is a *circular* definition, a fake definition, for it only defines in terms of itself, as though one defined a cat simply to be those creatures which have all cat characteristics.

It is, of course, perfectly logical and proper to work out a formal mathematical theory which describes the behavior of "things" which satisfy the formal postulates of the theory. If it turns out that such things do exist in experience, then the results of the theory will apply. If not, the mathematicians can say that it was nevertheless fun writing the poetry, even though it turned out to be nonsense verse.

Having gone through the reasoning of Chapter XI,

on the law of large numbers, we realize now the intimate and significant relation between the probabilities of events (as formally defined) and the frequency with which such events occur in long runs of trials.

The undeniable existence of such a relation has led mathematicians to attempt to develop theories which *start out* with a frequency definition of probabilities, rather than with a necessarily artificial "equally probable" definition.

But this turns out to be rough going also. You cannot speak accurately of the experimental value of a long-run frequency, for in mathematics the phrase "long-run" implies going on forever; and no one has ever, or can ever, perform an actual experiment which goes on forever. Therefore, some dodge must be introduced which in essence asks you to accept the "theoretical" long-range frequencies. But if you are going to swallow that, perhaps you would as soon swallow "equally probable events."

In many books on probability you will find a notation such as

$$P(E/F)$$

standing for "the probability P of the event E provided you know (or accept) F." Thus for a coin you might say

$$P(E/F) = 1/2$$

meaning that the probability for the event E (tossing a head) is 1/2 provided you accept F (heads and tails are equally probable) or F (the coin is "fair" so that ratio of heads to tails will in the long run tend to be equal).

You might say

$$P(E/F) = 1/2$$

meaning that the probability P of cutting a red card is 1/2 provided you know or accept F (all the 52 cards are present and the deck has been well shuffled).

But you should also write

$$P(E/F) \neq 1/2$$

if, in the same case of cutting a full deck and granting that it be well shuffled, F means "one or two of the red cards have sticky faces."

This notation emphasizes the necessity of recognizing the information on the basis of which you judge or adopt probability values.

There are important circumstances in which it is very tricky to estimate probabilities on the basis of the experience already in hand concerning frequencies. Suppose, for example, you know that someone had drawn 10 balls out of a jar that contains a mixture of colored balls, nothing whatsoever being known, prior to drawing the 10, about the distribution of various colors in the jar. Suppose of the 10 drawn, 3 are white and 7 black. What is the probability, on this evidence, that the next ball drawn will be white?

If you are ready to say that the answer is 3/10, then you better think a while longer. For suppose I substitute a new question: What is the probability that the next ball drawn will be red? Knowing *nothing* about the original mixture, are you prepared to say that the evidence of the 10 drawn justifies you in concluding that there is not a single red ball in the jar, so that the probability is zero?

You should not so conclude. On the contrary, you should say: This is a problem in "inverse probabilities." Can I apply Bayes' theorem (see page 308)? No, I cannot unless I am prepared to make some assumptions about the original distribution in the jar, and then to alter those views on the basis of the evidence in hand.

I do not want to do more here than to reawaken, near the end of this book, your interest in the basic concept of probability. It is a fascinating and complicated and deep subject.[1] The debate concerning it will, I am sure, go on forever. In the meantime, do not for one moment forget one massive fact about probability theory — it works.

Geometrical Probabilities

In all the discussion of this book we have been talking about n cases, where n is an integer. That is, we have been talking about a discrete and finite number of possible outcomes. But think about a straight line, say 10″ long. A point is going to be placed on this line at random — that is, in such a way that it is just as likely to land one place as another.

What is the probability that it will land exactly in the middle? The number of points on this line is infinite — or, more simply and accurately, as you subdivide the line into shorter and shorter segments the number of these segments increases without limit. Thus here we are not dealing with a finite number of cases. It isn't very sensible to ask for the probability that the point will land *exactly* in the center. For one thing, no one could ever decide

[1] I. G. Good, a mathematician with the Admiralty Research Laboratory in England, wrote a fine and stimulating article, "Kinds of Probability," in the February 20, 1959 issue of *Science*.

whether or not this had happened. It is, however, sensible to ask: What is the probability that it will land on a segment 0.01″ wide whose center is at the center? Clearly, the sensible answer to this question is 0.01/10 or 0.001.

A rather nice problem in geometric probabilities is this. Consider a line segment AB with a point located somewhere to the right of the center point, thus dividing the line segment into two pieces of length a and b, with b smaller than a. Choose a point C at random in the longer segment, and another point D at random in the shorter segment. These two points divide the line into three pieces AC, CD, and DB. What is the probability that a triangle can be formed out of these three pieces?

This is not trivially easy, and the answer, namely, $p = b/2a$, is very simple and elegant.

To indicate that one must be very careful when dealing with an infinite number of cases, we might note the classical question first discussed by the French mathematician L. F. Bertrand in 1889. Consider a random chord of a circle. What is the probability that it is shorter than a side of an inscribed equilateral triangle? But what is a "random chord?"

You must agree that it is completely indifferent what point X on the circumference is occupied by one end of the chord (for the circle is round, and no one point on it has any special location), and if you say that all angles between the chord and the line from X to the center are equally probable, then it easily turns out that the probability in question has the value 2/3.

But if you say that the way to get a "random

chord" is to locate its mid-point at random, then this mid-point must be more than a half radius from the center if the chord is to be shorter than a side of the inscribed equilateral triangle. And three-fourths of the area of the circle lies in the band reaching from the half-radius to the circumference. So on this basis the probability comes out to be 3/4.

There is another possible definition of a random chord, and that one leads to the intermediate value 1/2 for the probability. The point is that you have to agree about the "equally probable cases"; and when dealing with continuous variables and an "infinite number of cases," you have to watch your step.

Another classical problem of geometrical probability is Buffon's problem. Suppose a plane surface is ruled with parallel lines all a distance a apart (as the cracks on a hardwood floor). Throw on this plane, at random, a rod (match, needle, etc.) of length c, where c is less than a. What is the probability that the rod will cross one of the lines? The answer is $2c/\pi a$. Thus by throwing the rod a large number of times N and observing the number of times m the rod crosses the crack, you can calculate an "experimental" value of π as,

$$\frac{2cN}{am}$$

At first it seems almost magical that the answer involves the number π. But remember that π is connected with circles and with angles. If the center of the needle lands at some random spot, an end of the needle could land anywhere on a circle about that spot — for it is by the conditions of the problem as likely to have one orientation as another. Thus the circle comes into the problem, and π with it.

Over a hundred years ago a Professor Wolf of Frankfort threw a needle 5000 times; the needle was 36 mm. long and the plane was ruled with lines 45 mm. apart. He observed that the needle crossed a line 2532 times; from which one calculates 3.1596 as an approximation (with an error of only 0.6 per cent) to $\pi = 3.1416$.

About the turn of the century a Captain Fox is stated to have carried out 1120 trials with the result 3.1419; and still later a mathematician named Lazzarini carried out 3408 trials and obtained the value 3.1415929, which differs from the true value $\pi = 3.14159265$ only in the seventh decimal place! It is easy to guess that there was something a little fishy about S. Lazzarini's experiment, and this matter has recently been interestingly discussed.[2]

It Can't Be Chance!

Many of the discussions of this book should make you stop and think hard when someone says, "The probability of thus-and-so's happening by pure chance is only [say] 0.005. It is therefore wholly unreasonable to suppose that so improbable a thing has occurred, and hence you must conclude that chance was *not* operating, but that on the contrary there was some real *cause* back of the event." In fact, you know that when you get a hand at bridge (which certainly happened by chance), you have observed an event whose probability is far, far smaller than 0.005.

[2]"Puzzles and Paradoxes," by T. H. O'Beirne. *The New Scientist*, No. 238, 8 June 1961, p. 598.

As long ago as 1861 the German spectroscopist G. R. Kirchhoff said, "The probability that this coincidence is a mere work of chance is, therefore, considerably less than $(1/2)^{60}$ . . . . Hence this coincidence must be produced by some cause." He was examining sixty lines in the spectrum of iron; he compared these with sixty lines in the spectrum of the sun, and concluded that the coincidences were not accidental but that there must be iron in the sun. I think we all would agree with him, for if a coin came up heads sixty times in a row, I believe we would all conclude that there was some trick involved. A fake coin or a specially trained tosser would be much more credible than an honest sequence of sixty uninterrupted heads.

Notice, however, that what one should do is *weigh alternatives.* Seldom is it very meaningful to say, "I simply can't believe that." Usually one should say, "That seems to me much *harder* to believe *than* . . ."

Considerations of this sort are essential to the weighing of the evidence, largely produced by Professor J. B. Rhine[3] and his associates at Duke University, concerning extrasensory perception (ESP).

In the book cited there are described experiments which are interpreted to prove that a person can be "aware" of what is in another mind (telepathy); can "perceive" objects or events at a distance (clairvoyance) without any use of the senses of sight, hearing, touch, etc.; and indeed that the mind can directly, and without any other than a purely mental effort,

[3]See J. B. Rhine and J. G. Pratt, *Parapsychology* (Springfield, Illinois: Charles C Thomas, 1957).

have a material effect on a physical system (psychokinesis — for example, "willing" that one face of a die come up significantly more than one-sixth of the time).

The experiments in question involve a large number of trials, such as trials for naming cards as they are observed, one at a time, by some other person; these trials "succeed" a certain proportion of the time; and a calculation is then made of the probability that the successes are due to mere chance. When the probability, on a chance basis, turns out to be very small, the authors conclude that it could not have been "chance," but must therefore have been due to extrasensory perception.

In Chapter 9 of the book cited and speaking of the probability P for the occurrence of some strange event,[4] they say (page 172), "If P is sufficiently small the 'chance' explanation becomes unreasonable. . . ." Thus on the next page, calculating $P = 0.0005$ for a certain occurrence, they conclude that this is ". . . so unlikely that the chance hypothesis is not a reasonable explanation of the results."

It is not fair to argue against them that this is a completely isolated judgment, involving no comparison with alternatives. For $P = 0.0005$ is the probability of all possibilities as strange or stranger than what has in fact occurred; hence, 0.9995 is the alternative probability of something less strange. In one sense they are saying, the probability 0.0005 seems to be unacceptably small as compared with 0.9995.

[4]More accurately, P is usually the probability for events as strange as, or stranger than, the stated event.

But this comparison is, unhappily, not the one that is most meaningful to many persons. On the one hand, we are asked to accept an interpretation that destroys the most fundamental ideas and principles on which modern science has been based; we are asked to give up the irreversibility of time, to accept an effect that shows no decay with distance and hence involves "communication" without energy being involved; asked to believe in an "effect" that depends on no known quantities and for which no explanation has been offered, to credit phenomena which are subject to decline or disappearance for unexplained and unexplainable reasons. On the other hand, we are asked not to believe that a highly improbable chance result has occurred. All I can say is, I find this a very tough pair of alternatives.

The Rhine ESP results could be explained on the grounds of selection or falsification of data. Having complete confidence in the scientific competence and personal integrity of Professor Rhine, I find this explanation unacceptable to me. In any very long probability experiment there will occur highly remarkable runs of luck — as in the 28 recorded repetitions of one color at Monte Carlo, or the long runs of "passes" at craps (see page 281). But I know of no analysis of Rhine data, based on such considerations, that makes it reasonable to believe that their success can be explained in this way.

On various occasions Professor Rhine, or his
supporters, have pasted together, so to speak, one
after another of their experiments to produce one 1
grand over-all experiment whose probability, on the
assumption of chance, then turns out to be 10 with a

very large negative exponent. But this procedure does not seem to me impressive. In a certain sense this is (although perfectly honest and open) a selection of data. It is, for example, only human that they do not include, in this compounding, long runs made *after* a certain subject had gone into an ESP decline. And it is very easy — indeed almost trivial — to compound unlikely events to produce a miracle. The probability that a given hand will contain thirteen specified cards (say thirteen spades) is of the order of 10^{-12}. I have heard of its happening in n cases. The compound probability that all these thirteen persons hold all spades is 10^{-12n}. So what?

As I have said elsewhere, I find this a subject that is so intellectually uncomfortable as to be almost painful. I end by concluding that I cannot explain away Professor Rhine's evidence, and that I also cannot accept his interpretation.

The Surprising Stability of Statistical Results

One of the most striking and fundamental things about probability theory is that it leads to an understanding of the otherwise strange fact that events which are individually capricious and unpredictable can, when treated *en masse*, lead to very stable *average* performances.

Perhaps two rather unusual examples of this are worth noting. The circumstances which result in a dog's biting a person seriously enough so that the matter gets reported to the health authorities would seem to be complex and unpredictable indeed. In New York City, in the year 1955, there were, on the average, 75.3 reports per day to the Department of

Health of bitings of people. In 1956 the corresponding number was 73.6. In 1957 it was 73.5. In 1957 and 1958 the figures were 74.5 and 72.6.

According to a British official report, for the decade 1931–40, the average number of murders in England per million of population was 3.2. Skipping the abnormal decade of World War II, the same average over the 1951–60 period was 3.3. This result, of course, carries with it implications about social stability in England.

The Subtlety of Probabilistic Reasoning

We have seen, at various points in this book, that probability theory is no subject to approach with a lazy or fuzzy mind. The example concerning the chord of a circle, given above, illustrates that you must be constantly on your mental toes. Near the end of this book I want to warn you that this is no blunt instrument with which you may go about indiscriminately blipping problems over the head. You are offered a delicate and precise tool, and it must be used with care.

You know by now what to think of the doctor (an imaginary one invented[5] by the mathematician G. Polya) who comforted his patient with the remark, "You have a very serious disease. Of ten persons who get this disease only one survives. But do not worry. It is lucky you came to me, for I have recently had nine patients with this disease and they all died of it."

If we have time for just one more probability

[5]*Patterns of Plausible Reasoning*, by G. Polya (Princeton, New Jersey: Princeton University Press, 1954), Vol. II, p. 100.

paradox, consider the problem of the second ace.[6] I am going to describe this paradox in the language Mr. Gardner uses, and then later ask if this language is correct.

A hand is dealt, and you state to the other three players, "I have an ace." What is the probability, as calculated, of course, by someone other than you (for you know what you have) that you also have *another* ace? This, says Mr. Gardner, is 5359/14498, which is about 0.37.

A few hands later you announce, "I have the Ace of Spades." What now is the probability that you also have another ace?

Surprisingly enough, Mr. Gardner says, the answer now is 11686/20825, or about 0.56. Why does the knowledge of *which* ace affect the answer?

Mr. Gardner goes on to say that you can analyze this situation by considering the much simpler but similar problem of dealing a hand of *two* cards from a "deck" consisting of only four cards, namely the Ace of Spades, Ace of Hearts, Jack of Diamonds, and two of Clubs. There are only 6 possible two-card hands from this deck. Five of them have at least one ace and in only one of these five is there a second ace. Three of them contain the Ace of Spades, and one of these three contains a second ace. So in this illustration the two answers would be 1/5 and 1/3! But is this reasoning of Mr. Gardner's really sound? Are the five hands (in the first case) and the three hands (in the second case) *equally probable?* Is this similar to the old problem of the three chests?

[6]See *The Scientific American Book of Mathematical Puzzles and Diversions*, by Martin Gardner. (New York: Simon and Schuster, 1959), p. 49.

The Modern Reign of Probability

Over thirty years ago I wrote a paper with the title "The Reign of Probability." There was a large amount of impressive evidence at that time for the importance of the theory of probability. But the case has so substantially strengthened, over the intervening years, that I must use, for this closing section, an up-to-date title.

It would be easy to write a book devoted solely to the task of setting forth the areas of modern experience within which the theory of probability plays not just *some* role, but often a completely fundamental one. So I will have to be content with nothing much more than a list of topics—the list surely incomplete and the explanations not much more than hints which I hope will tease your curiosity.

As a first point, note that to a very great extent science proceeds by making *measurements* and then fitting these measurements into a unifying and simplifying theory. In fact, it has been rather popular for scientists to remark that you don't really understand anything until you know how to measure it.[7] The more carefully and more precisely you measure any quantity, the more certain it is that if you repeat the process you will get a different answer. If asked to guess to the nearest foot the length of a bar about two feet long, most of the guesses would coincide. But if ten different experts

[7]This remark is of limited usefulness. Science has probably never measured anything as frequently or as accurately as intervals of *time;* but I am not aware that this persistence has at all increased an understanding of what "time" really is.

each measured its length ten separate times to the nearest hundred-millionth of an inch, it is entirely possible that the result would be nearly one hundred different answers.

In other words, measurement, in this imperfect world of ours, is subject to error. *You can't make progress toward precision unless you have some way of dealing with error.*

The theory of probability — in its special branch known as the theory of errors — furnishes this method. It is not possible, in my judgment, to over-emphasize the importance of this.

Second, knowledge would be impossible (and indeed unimaginable) unless men could *communicate,* one with another. As thinking men always have realized, language is an imperfect instrument; it is hard to say exactly what you mean, and still harder to try to insure that your listener or reader will really know what you are trying to convey. And if, in turn, you have difficulty in understanding him, just how does anyone ever decide whether or not an idea has or has not been conveyed, intact and unaltered, from your brain to that of another person?

Only in the last quarter century has it been clearly recognized that communication is essentially a probabilistic process. The fact is that you can never be *certain* about communication. The best you can do is to *increase the probability* that a message is received without error. The mathematical theory of communication and mathematical information theory have been developed vigorously during the last few years, and useful application has been found to a very wide range of subjects in biology, medicine, psychology, etc., in addition to the more expected applications in telephony, television, linguistics, etc. *Language,* we have learned, is a "stochastic process," to use the popular phrase, which simply means "a procedure which is governed by the laws of probability."

Third, as I hinted in the opening chapter, we are recognizing more and more strongly that *probabilistic reasoning* is the sort of reasoning which most of us have to do most of the time. I do not in the least mean to imply — for that would be ridiculous — that most persons, faced with a problem, stop and invoke theorems out of a probability textbook. I *do* mean that some knowledge of the nature of probabilistic reasoning, of its power and of its pitfalls, would assist almost every person to cope better with the job — which is the normal and usual one — of drawing conclusions from incomplete or imperfect evidence.

If one can argue reasonably that it is not possible to *measure, think,* or *communicate* without some aid from probability theory, it would be formally correct to rest the case at this point. But I want to go on, just a bit further, to indicate more specifically how probability theory is useful — or rather that it is inescapable — in science.

The material universe, including all living creatures, is composed — as the defining adjective implies — of *matter.* There may be (I happen to think that there are) aspects of life which can never be analyzed by scientific techniques; but up to date the progress of not only the physical sciences but the biological sciences too has depended on observational and experimental investigation of the properties of, and the behavior of, matter. It has been found neither necessary nor possible to suppose that living matter follows any laws peculiar to it; nor has it been necessary to suppose that living matter, *as the scientist analyzes it*, contains anything that a physicist or a chemist would consider mysterious. So much of modern progress in biology is at the level of submicroscopic phenomena where the biochemist and biophysicist furnish most of the techniques that it has now become commonplace to talk of "molecular biology."

Thus the basic laws of physical matter are nothing less than the basic laws of *all* matter. And when you get right down to brass tacks, down to the level of detail which is, at least at present, as far down as man's mind can go, what kinds of laws are these?

The behavior of molecules, atoms, elementary particles, such as the electron, proton, neutron, meson, etc., is all analyzed in the subject called quantum theory. And *the basic laws of that theory are probabilistic.*

In quantum theory one always wishes to determine what will be the "state" of the system under study (it might be an atom, or a group of molecules, or even the nucleus of an atom). This means, roughly, where will the component parts be, how will they be moving, what energy will they have, etc.?

Quantum theory characteristically proceeds by setting up equations which, when solved, result in what is called the "Schrodinger wave function." And from this function, by a little mathematical manipulation, you find out — what? You find out the *probabilities* that the component parts of your system are here or there, are moving thus or so, have this or some other energy.

Suppose you have your mental eye on an electron

and want to know how it is moving and where it is. If you try to squeeze nature into a corner by inventing more and more precise ways of finding out *where* the electron is, nature escapes and makes it more and more vague *how the electron is moving.* If you pin down the velocity more and more accurately, then you get less and less definite about where the electron is. You simply have to be content with probability estimates.

There have been a few leading scientists—Einstein is the chief example—who have regretted and resented this situation, and viewed it as a symptom of temporary fault in physical theories. Einstein did not feel comfortable about a probabilistic basis for all science and once remarked that, in his judgment, "God does not play dice." The large majority of scientists, including many truly great ones such as Bohr and Heisenberg, accept the modern reign of probability.

If the basic laws of the most elementary physical phenomena are probabilistic, then there is not too much point in going on to mention that probability is essential for many, many special fields of physics. As for the astronomers, not only do they use the theory to refine the accuracy and reliability of all their numerical observations, not only is it essential for all the considerations of astrophysics, but during recent years the most powerful statistical techniques have been applied to problems of star distributions and other similar questions bearing on the structure of the universe.

The space scientists[8] "are now busy calculating

[8]*Industrial Bulletin,* Arthur D. Little, Inc., March 1961.

the chances that a space ship and its crew can survive an interplanetary voyage." This calculation includes, for example, the probability of a chance collision with one or more meteors.

The chemists, who nowadays are often hard to distinguish from the physicists, are equally concerned with quantum dynamics, and hence with probability. All aspects of thermodynamics, kinetic gas theory, and the multitude of applications of statistical mechanics in chemistry, involve probabilistic considerations. When electrons impact on carbon monoxide, or nitric oxide, or on oxygen, negative oxygen ions may be formed, but what modern chemistry can tell you is the *probability* that this will occur. The centrally important theory of the rates at which chemical reactions occur is based on probability considerations.

A search for buried treasure is clearly a chancy business. You would expect exploring geologists and geophysicists to think in probability terms. My lifelong friend Professor Louis B. Slichter, Director of the Institute of Geophysics at the University of California at Los Angeles, has written an article about search for ore deposits.[9] The first section of this article is titled "The Gambler," and discusses the classic problem of the gambler's ruin. The next two sections bear the titles "Statistical Analysis of Ore Body Occurrences," and "Probability of Discovery."

Indeed, *The New Yorker* (March 18, 1961) in a series of articles on Texas, quoted a saying of the oilman Ted Weiner of Fort Worth, "The oil business

[9]"The Need of a New Philosophy of Prospecting," *Mining Engineering*, June 1960, pp. 1–7.

is not only a game but — what is even better suited to the American taste — a game of chance. It's just like running a dice table — one that's honest, open, and all above board."

The article goes on:

> What are the chances of success? The mathematical probabilities fluctuate in accordance with, for one thing, the location of the well. As might be expected, the odds are longest on *discovery wells* (those drilled where oil has not previously been found) and shortest on *development wells* (those put down on acreage where oil is already being produced). According to averages compiled by the industry, the chance of drilling a successful discovery well are one in eight, and of drilling a successful development well, three in four. The probabilities also vary from one operator to another . . . a number of experienced operators . . . have a fairly consistent average of about one in five (discovery type).
>
> To be within the margin of profitability, according to the industry's estimate, a discovery well must bring in a field that will produce a million barrels of oil. The chance of finding such a field is one in forty-three. The chance of finding a ten-million-barrel field are one in two hundred forty-three, and of finding a fifty-million-barrel field, one in nine hundred and sixty-seven.
>
> . . . Of the thousands of wells drilled in this county since the beginning of the century, less than fifty have brought in a hundred-million-barrel field.

Clearly Lady Luck still plays games at times, and out of doors as well as in.

Biology is deeply concerned with probability; genetics very frequently uses probabilistic ideas and calculations. When a male sperm cell unites with a female egg cell to form the beginning of a new creature, plant or mouse or man, there occurs a shuffling of the genes from the maternal line and those from the paternal line, so that it is a matter of chance what particular set of pairs of genes the new creature has, each of his pairs consisting of one chosen by chance from the relevant pair of the male parent, and one chosen similarly from the relevant pair of the maternal parent.

Suppose a certain pair of genes controls *color*—say the green or yellow seed color in Mendel's original experiment with sweet peas. The plant in whose cells are two "yellow genes" produces yellow seeds. Those with two "green genes" produce green seeds. Those with one yellow gene and one green gene produce yellow seeds; the yellow gene is *dominant*, as geneticists say, having its way even though the other gene votes for green.

When a purebred yellow parent (that is, one whose cells all contain two yellow genes) is crossed with a purebred green parent (whose cells all have two green genes), the first generation result is offspring which necessarily have one yellow and one green gene. They all produce yellow seeds.

But if you now produce a second generation, what will they turn out to be? Each second generation progeny receives one gene from one parent (which has an equal chance of being a green or a yellow gene), and one from the other parent (which also has an even chance of being green or yellow).

This is precisely analogous to the case of two persons tossing a coin, once each. There is an equal chance of getting

First Person	*Second Person*
H (Yellow)	H (Yellow)
H (Yellow)	T (Green)
T (Green)	H (Yellow)
T (Green)	T (Green)

But since in the case of the sweet pea plants the yellow influence outweighs the green influence, in only one-quarter of the instances (the case listed last) will the seeds be green, whereas in three-quarters (the first, second, and third cases) the seeds will be yellow.

And if you now made additional crosses within the four groups, the fourth one would "breed true" and continue to produce green seeds. The three-fourths composed of the first, second, and third cases all look alike, as regards their seed color; but you know that one-third of them will breed true yellow every time, whereas for the other two-thirds there is a calculable probability (indeed it is one chance in three) that the chance shuffling will bring together two green genes and produce green seeds.

Or consider a classic experiment carried out by the German botanist Karl Erich Correns, who was one of those who "rediscovered" Mendel's work in 1900 after it had been unnoticed for thirty-five years, buried in an obscure journal.

A plant with a white flower is crossed with a closely related plant (different species of the same genus) with a red flower. The first generation offspring all have pink flowers (one white gene and one red gene). Of the second generation some are red, about the same number white, and more are pink.

Fig. 49 shows the actual numbers in Correns' experiment, which involved 565 plants of the third generation.

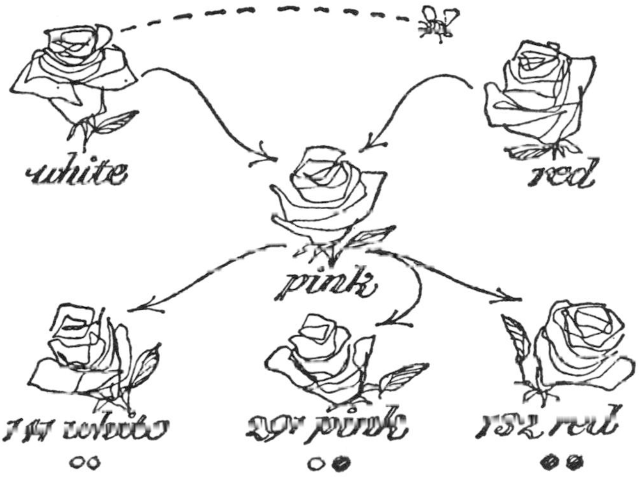

FIG. 49

Notice that 141:292:132 almost as 1:2:1. Can you not readily supply the probability argument for these ratios?

Still considering probability and genetics, it is interesting to note that the numbers of chromosome interchanges per cell produced by a given dosage of X-radiation follow the Poisson distribution law.

These very simple, but historically classic examples, give a faint indication of the way in which probability calculations enter into genetics. Books and books have been written on this subject.

One of the most central problems of biology is concerned with the fact that of three small, nearly spherical blobs of protoplasm, one will develop into an elephant, say, with legs and a trunk, one into a bird with wings, another into a snake or a man. A young English genius named Alan M. Turing,[10] originally trained as a mathematician, wrote a paper[11] which is regarded as most imaginative and important. In a critique of this paper Dr. J. W. S. Pringle, formerly of Cambridge University and now of Oxford, has said: "It follows from Turing's arguments that in a system which is initially completely homogeneous, the positions at which local concentrations will appear is indeterminate *in the sense that it is 'caused' by chance fluctuations. . . .*" (italics mine). In other words, Lady Luck is at work here, inside an egg cell in its earliest stages, affecting the initial decisions that determine how this cell develops.

The whole area of medical research is one that cries out for more active and more general use of probability thinking. The need for statistical competence in the interpretation of results is widely recognized, although not always expertly served. The need for really competent advice on the planning of experimentation is even greater, and as far as I can judge, less well served. In many instances (say those connected with severe but

[10]Who also wrote exceedingly important papers about computing machines, their ability to "think," and their ultimate limitations.

[11]"The Chemical Basis of Morphogenesis," *Philosophical Transactions* of the Royal Society, Series B, No. 641, 14 August 1952.

infrequent human disorders) it is of the greatest importance to squeeze the last drop of dependable significance out of the necessarily meager experimental evidence. When investigators are trying out new drugs, or screening chemicals or biologicals for possible effectiveness against cancer, it is of the clearest importance to use the most refined, precise, and dependable procedures for reaching decisions and for judging the dependability of those decisions. All this can be furnished by probability and statistics, and by nothing else.

It seems highly probable that all fields of the social sciences will, as they become more firmly grounded, make increasing use of probability theory. Questions concerning legal judgments have been considered, by probabilists, ever since the anonymous paper "A Calculation on the Credibility of Human Testimony," which appeared in the *Philosophical Transactions* in England in 1699. Not quite a century later, the Frenchman Condorcet wrote his treatise which, in translation, is titled "Essay Concerning the Application of Probability Analysis to Plurality Decisions."

I could refer you to a recent article[12] which illustrates how currently interesting and important are the applications of probability to the machinery of the democratic process.

Or if you do not mind ending on a somewhat comic

[12]"Some Statistical Problems of Majority Voting," by Professor L. S. Penrose, *The New Scientist*, No. 224, 2 March 1961. The preface to a current best seller, *My Life in Court*, by Louis Nizer (New York: Doubleday & Company, 1961), contains very interesting proof of the fact that a skilled trial lawyer finds probability thinking not just useful, but necessary.

note, consider the problem "If A, B, C, and D each speak the truth once in three times (independently), and if A affirms that B denies that C declares that D is a liar, what is the probability that D was speaking the truth?"

This problem, sometimes called "the truth about four liars," is by no means an uncomplicated one. Sir Arthur Eddington used it[13] in a book he wrote in 1935, but gave a "solution" which he later confessed was inadequate. A careful discussion of this problem has been published recently.[14] It is necessary, before the problem becomes really meaningful, to add some further information. Eddington's answer of 25/71 holds only if one adds a condition which is rather ridiculous in character. With what seems to be the most sensible additions, the answer is 13/41.

Lady Luck and the Future

Probability theory and statistics have now become an important and in fact a necessary part of our lives. Great industries, such as all forms of insurance, are largely dependent upon the laws of probability. Every aspect of science that involves measurement is inescapably concerned with that branch of probability known as the theory of errors. The physical world has been found to be essentially probabilistic in nature. Basic aspects of biology are probabilistic. Every appeal to experience that utilizes a sampling procedure depends, for its interpretation, upon

[13]*New Pathways in Science* (New York: The Macmillan Company, 1935), p. 121.

[14]*The New Scientist*, No. 234, 11 May 1961, p. 330.

statistical theory. Very many of the judgments and decisions which we all have to make every day are based upon a conscious or intuitive — and doubtless chiefly the latter — weighing of probabilities. Even courage, as Socrates pointed out, is "the knowledge of the grounds of fear and hope," and thus depends upon our probabilistic estimates of the risks involved in our estimates of fears and hopes.

But in spite of this importance, those in charge of educational affairs have not as yet properly recognized the universal importance of probabilities and statistics. There is no doubt whatsoever that mathematicians, over the years ahead, will develop still deeper, still more general, and still more powerful theorems in probability and statistics. It is profoundly to be hoped that there will be educational progress along with this scientific progress, that the elements of probability theory will be included in high school courses, and that this exceedingly interesting and practical material will replace various outmoded mathematical subjects of purely historical and highly specialized interest. Still more is it to be hoped that colleges will see to it that all those who expect to enter the physical, biological, medical, or social sciences have a grounding in probability or statistics. These subjects have proved their worth, and have come of age. Lady Luck has become a lady.

INDEX

A CATALOG OF SELECTED

DOVER BOOKS

IN ALL FIELDS OF INTEREST

THE BOOK OF BEASTS: Being a Translation from a Latin Bestiary of the Twelfth Century, T. H. White. Wonderful catalog real and fanciful beasts: manticore, griffin, phoenix, amphivius, jaculus, many more. White's witty erudite commentary on scientific, historical aspects. Fascinating glimpse of medieval mind. Illustrated. 296pp. 5⅜ × 8¼. (Available in U.S. only) 24609-4 Pa. $6.95

FRANK LLOYD WRIGHT: ARCHITECTURE AND NATURE With 160 Illustrations, Donald Hoffmann. Profusely illustrated study of influence of nature—especially prairie—on Wright's designs for Fallingwater, Robie House, Guggenheim Museum, other masterpieces. 96pp. 9¼ × 10¾. 25098-9 Pa. $8.95

FRANK LLOYD WRIGHT'S FALLINGWATER, Donald Hoffmann. Wright's famous waterfall house: planning and construction of organic idea. History of site, owners, Wright's personal involvement. Photographs of various stages of building. Preface by Edgar Kaufmann, Jr. 100 illustrations. 112pp. 9¼ × 10. 23671-4 Pa. $8.95

YEARS WITH FRANK LLOYD WRIGHT: Apprentice to Genius, Edgar Tafel. Insightful memoir by a former apprentice presents a revealing portrait of Wright the man, the inspired teacher, the greatest American architect. 372 black-and-white illustrations. Preface. Index. vi + 228pp. 8¼ × 11. 24801-1 Pa. $10.95

THE STORY OF KING ARTHUR AND HIS KNIGHTS, Howard Pyle. Enchanting version of King Arthur fable has delighted generations with imaginative narratives of exciting adventures and unforgettable illustrations by the author. 41 illustrations. xviii + 313pp. 6⅛ × 9¼. 21445-1 Pa. $6.95

THE GODS OF THE EGYPTIANS, E. A. Wallis Budge. Thorough coverage of numerous gods of ancient Egypt by foremost Egyptologist. Information on evolution of cults, rites and gods; the cult of Osiris; the Book of the Dead and its rites; the sacred animals and birds; Heaven and Hell; and more. 956pp. 6⅛ × 9¼. 22055-9, 22056-7 Pa., Two-vol. set $21.90

A THEOLOGICO-POLITICAL TREATISE, Benedict Spinoza. Also contains unfinished *Political Treatise*. Great classic on religious liberty, theory of government on common consent. R. Elwes translation. Total of 421pp. 5⅜ × 8½. 20249-6 Pa. $7.95

INCIDENTS OF TRAVEL IN CENTRAL AMERICA, CHIAPAS, AND YUCATAN, John L. Stephens. Almost single-handed discovery of Maya culture; exploration of ruined cities, monuments, temples; customs of Indians. 115 drawings. 892pp. 5⅜ × 8½. 22404-X, 22405-8 Pa., Two-vol. set $15.90

LOS CAPRICHOS, Francisco Goya. 80 plates of wild, grotesque monsters and caricatures. Prado manuscript included. 183pp. 6⅝ × 9⅝. 22384-1 Pa. $5.95

AUTOBIOGRAPHY: The Story of My Experiments with Truth, Mohandas K. Gandhi. Not hagiography, but Gandhi in his own words. Boyhood, legal studies, purification, the growth of the Satyagraha (nonviolent protest) movement. Critical, inspiring work of the man who freed India. 480pp. 5⅜ × 8½. (Available in U.S. only) 24593-4 Pa. $6.95

ILLUSTRATED DICTIONARY OF HISTORIC ARCHITECTURE, edited by Cyril M. Harris. Extraordinary compendium of clear, concise definitions for over 5,000 important architectural terms complemented by over 2,000 line drawings. Covers full spectrum of architecture from ancient ruins to 20th-century Modernism. Preface. 592pp. 7½ × 9⅜. 24444-X Pa. $15.95

THE NIGHT BEFORE CHRISTMAS, Clement Moore. Full text, and woodcuts from original 1848 book. Also critical, historical material. 19 illustrations. 40pp. 4⅝ × 6. 22797-9 Pa. $2.50

THE LESSON OF JAPANESE ARCHITECTURE: 165 Photographs, Jiro Harada. Memorable gallery of 165 photographs taken in the 1930's of exquisite Japanese homes of the well-to-do and historic buildings. 13 line diagrams. 192pp. 8⅜ × 11¼. 24778-3 Pa. $10.95

THE AUTOBIOGRAPHY OF CHARLES DARWIN AND SELECTED LETTERS, edited by Francis Darwin. The fascinating life of eccentric genius composed of an intimate memoir by Darwin (intended for his children); commentary by his son, Francis; hundreds of fragments from notebooks, journals, papers; and letters to and from Lyell, Hooker, Huxley, Wallace and Henslow. xi + 365pp. 5⅜ × 8. 20479-0 Pa. $6.95

WONDERS OF THE SKY: Observing Rainbows, Comets, Eclipses, the Stars and Other Phenomena, Fred Schaaf. Charming, easy-to-read poetic guide to all manner of celestial events visible to the naked eye. Mock suns, glories, Belt of Venus, more. Illustrated. 299pp. 5¼ × 8¼. 24402-4 Pa. $7.95

BURNHAM'S CELESTIAL HANDBOOK, Robert Burnham, Jr. Thorough guide to the stars beyond our solar system. Exhaustive treatment. Alphabetical by constellation: Andromeda to Cetus in Vol. 1; Chamaeleon to Orion in Vol. 2; and Pavo to Vulpecula in Vol. 3. Hundreds of illustrations. Index in Vol. 3. 2,000pp. 6⅛ × 9¼. 23567-X, 23568-8, 23673-0 Pa., Three-vol. set $41.85

STAR NAMES: Their Lore and Meaning, Richard Hinckley Allen. Fascinating history of names various cultures have given to constellations and literary and folkloristic uses that have been made of stars. Indexes to subjects. Arabic and Greek names. Biblical references. Bibliography. 563pp. 5⅜ × 8½. 21079-0 Pa. $8.95

THIRTY YEARS THAT SHOOK PHYSICS: The Story of Quantum Theory, George Gamow. Lucid, accessible introduction to influential theory of energy and matter. Careful explanations of Dirac's anti-particles, Bohr's model of the atom, much more. 12 plates. Numerous drawings. 240pp. 5⅜ × 8½. 24895-X Pa. $5.95

CHINESE DOMESTIC FURNITURE IN PHOTOGRAPHS AND MEASURED DRAWINGS, Gustav Ecke. A rare volume, now affordably priced for antique collectors, furniture buffs and art historians. Detailed review of styles ranging from early Shang to late Ming. Unabridged republication. 161 black-and-white drawings, photos. Total of 224pp. 8⅜ × 11¼. (Available in U.S. only) 25171-3 Pa. $13.95

VINCENT VAN GOGH: A Biography, Julius Meier-Graefe. Dynamic, penetrating study of artist's life, relationship with brother, Theo, painting techniques, travels, more. Readable, engrossing. 160pp. 5⅜ × 8½. (Available in U.S. only) 25253-1 Pa. $4.95

SUNDIALS, Albert Waugh. Far and away the best, most thorough coverage of ideas, mathematics concerned, types, construction, adjusting anywhere. Over 100 illustrations. 230pp. 5⅜ × 8½. 22947-5 Pa. $5.95

PICTURE HISTORY OF THE NORMANDIE: With 190 Illustrations, Frank O. Braynard. Full story of legendary French ocean liner: Art Deco interiors, design innovations, furnishings, celebrities, maiden voyage, tragic fire, much more. Extensive text. 144pp. 8⅜ × 11¾. 25257-4 Pa. $10.95

THE FIRST AMERICAN COOKBOOK: A Facsimile of "American Cookery," 1796, Amelia Simmons. Facsimile of the first American-written cookbook published in the United States contains authentic recipes for colonial favorites—pumpkin pudding, winter squash pudding, spruce beer, Indian slapjacks, and more. Introductory Essay and Glossary of colonial cooking terms. 80pp. 5⅜ × 8½. 24710-4 Pa. $3.50

101 PUZZLES IN THOUGHT AND LOGIC, C. R. Wylie, Jr. Solve murders and robberies, find out which fishermen are liars, how a blind man could possibly identify a color—purely by your own reasoning! 107pp. 5⅜ × 8½. 20367-0 Pa. $2.50

ANCIENT EGYPTIAN MYTHS AND LEGENDS, Lewis Spence. Examines animism, totemism, fetishism, creation myths, deities, alchemy, art and magic, other topics. Over 50 illustrations. 432pp. 5⅜ × 8½. 26525-0 Pa. $8.95

ANTHROPOLOGY AND MODERN LIFE, Franz Boas. Great anthropologist's classic treatise on race and culture. Introduction by Ruth Bunzel. Only inexpensive paperback edition. 255pp. 5⅜ × 8½. 25245-0 Pa. $6.95

THE TALE OF PETER RABBIT, Beatrix Potter. The inimitable Peter's terrifying adventure in Mr. McGregor's garden, with all 27 wonderful, full-color Potter illustrations. 55pp. 4¼ × 5½. (Available in U.S. only) 22827-4 Pa. $1.75

THREE PROPHETIC SCIENCE FICTION NOVELS, H. G. Wells. *When the Sleeper Wakes, A Story of the Days to Come* and *The Time Machine* (full version). 335pp. 5⅜ × 8½. (Available in U.S. only) 20605-X Pa. $6.95

APICIUS COOKERY AND DINING IN IMPERIAL ROME, edited and translated by Joseph Dommers Vehling. Oldest known cookbook in existence offers readers a clear picture of what foods Romans ate, how they prepared them, etc. 49 illustrations. 301pp. 6⅛ × 9¼. 23563-7 Pa. $7.95

SHAKESPEARE LEXICON AND QUOTATION DICTIONARY, Alexander Schmidt. Full definitions, locations, shades of meaning of every word in plays and poems. More than 50,000 exact quotations. 1,485pp. 6½ × 9¼. 22726-X, 22727-8 Pa., Two-vol. set $31.90

THE WORLD'S GREAT SPEECHES, edited by Lewis Copeland and Lawrence W. Lamm. Vast collection of 278 speeches from Greeks to 1970. Powerful and effective models; unique look at history. 842pp. 5⅜ × 8½. 20468-5 Pa. $12.95

A CONCISE HISTORY OF PHOTOGRAPHY: Third Revised Edition, Helmut Gernsheim. Best one-volume history—camera obscura, photochemistry, daguerreotypes, evolution of cameras, film, more. Also artistic aspects—landscape, portraits, fine art, etc. 281 black-and-white photographs. 26 in color. 176pp. 8⅜ × 11¼. 25128-4 Pa. $13.95

THE DORÉ BIBLE ILLUSTRATIONS, Gustave Doré. 241 detailed plates from the Bible: the Creation scenes, Adam and Eve, Flood, Babylon, battle sequences, life of Jesus, etc. Each plate is accompanied by the verses from the King James version of the Bible. 241pp. 9 × 12. 23004-X Pa. $9.95

WANDERINGS IN WEST AFRICA, Richard F. Burton. Great Victorian scholar/adventurer's invaluable descriptions of African tribal rituals, fetishism, culture, art, much more. Fascinating 19th-century account. 624pp. 5⅜ × 8½. 26890-X Pa. $12.95

FLATLAND, E. A. Abbott. Intriguing and enormously popular science-fiction classic explores the complexities of trying to survive as a two-dimensional being in a three-dimensional world. Amusingly illustrated by the author. 16 illustrations. 103pp. 5⅜ × 8½. 20001-9 Pa. $2.50

THE HISTORY OF THE LEWIS AND CLARK EXPEDITION, Meriwether Lewis and William Clark, edited by Elliott Coues. Classic edition of Lewis and Clark's day-by-day journals that later became the basis for U.S. claims to Oregon and the West. Accurate and invaluable geographical, botanical, biological, meteorological and anthropological material. Total of 1,508pp. 5⅜ × 8½.
21268-8, 21269-6, 21270-X Pa. Three-vol. set $26.85

LANGUAGE, TRUTH AND LOGIC, Alfred J. Ayer. Famous, clear introduction to Vienna, Cambridge schools of Logical Positivism. Role of philosophy, elimination of metaphysics, nature of analysis, etc. 160pp. 5⅜ × 8½. (Available in U.S. and Canada only) 20010-8 Pa. $3.95

MATHEMATICS FOR THE NONMATHEMATICIAN, Morris Kline. Detailed, college-level treatment of mathematics in cultural and historical context, with numerous exercises. For liberal arts students. Preface. Recommended Reading Lists. Tables. Index. Numerous black-and-white figures. xvi + 641pp. 5⅜ × 8½.
24823-2 Pa. $11.95

HANDBOOK OF PICTORIAL SYMBOLS, Rudolph Modley. 3,250 signs and symbols, many systems in full; official or heavy commercial use. Arranged by subject. Most in Pictorial Archive series. 143pp. 8⅛ × 11. 23357-X Pa. $6.95

INCIDENTS OF TRAVEL IN YUCATAN, John L. Stephens. Classic (1843) exploration of jungles of Yucatan, looking for evidences of Maya civilization. Travel adventures, Mexican and Indian culture, etc. Total of 669pp. 5⅜ × 8½.
20926-1, 20927-X Pa., Two-vol. set $11.90

DEGAS: An Intimate Portrait, Ambroise Vollard. Charming, anecdotal memoir by famous art dealer of one of the greatest 19th-century French painters. 14 black-and-white illustrations. Introduction by Harold L. Van Doren. 96pp. 5⅜ × 8½.
25131-4 Pa. $4.95

PERSONAL NARRATIVE OF A PILGRIMAGE TO ALMANDINAH AND MECCAH, Richard Burton. Great travel classic by remarkably colorful personality. Burton, disguised as a Moroccan, visited sacred shrines of Islam, narrowly escaping death. 47 illustrations. 959pp. 5⅜ × 8½. 21217-3, 21218-1 Pa., Two-vol. set $19.90

PHRASE AND WORD ORIGINS, A. H. Holt. Entertaining, reliable, modern study of more than 1,200 colorful words, phrases, origins and histories. Much unexpected information. 254pp. 5⅜ × 8½. 20758-7 Pa. $5.95

THE RED THUMB MARK, R. Austin Freeman. In this first Dr. Thorndyke case, the great scientific detective draws fascinating conclusions from the nature of a single fingerprint. Exciting story, authentic science. 320pp. 5⅜ × 8½. (Available in U.S. only) 25210-8 Pa. $6.95

AN EGYPTIAN HIEROGLYPHIC DICTIONARY, E. A. Wallis Budge. Monumental work containing about 25,000 words or terms that occur in texts ranging from 3000 B.C. to 600 A.D. Each entry consists of a transliteration of the word, the word in hieroglyphs, and the meaning in English. 1,314pp. 6⅞ × 10.
23615-3, 23616-1 Pa., Two-vol. set $35.90

THE COMPLEAT STRATEGYST: Being a Primer on the Theory of Games of Strategy, J. D. Williams. Highly entertaining classic describes, with many illustrated examples, how to select best strategies in conflict situations. Prefaces. Appendices. xvi + 268pp. 5⅜ × 8½. 25101-2 Pa. $6.95

THE ROAD TO OZ, L. Frank Baum. Dorothy meets the Shaggy Man, little Button-Bright and the Rainbow's beautiful daughter in this delightful trip to the magical Land of Oz. 272pp. 5⅜ × 8. 25208-6 Pa. $5.95

POINT AND LINE TO PLANE, Wassily Kandinsky. Seminal exposition of role of point, line, other elements in non-objective painting. Essential to understanding 20th-century art. 127 illustrations. 192pp. 6½ × 9¼. 23808-3 Pa. $5.95

LADY ANNA, Anthony Trollope. Moving chronicle of Countess Lovel's bitter struggle to win for herself and daughter Anna their rightful rank and fortune—perhaps at cost of sanity itself. 384pp. 5⅜ × 8½. 24669-8 Pa. $8.95

EGYPTIAN MAGIC, E. A. Wallis Budge. Sums up all that is known about magic in Ancient Egypt: the role of magic in controlling the gods, powerful amulets that warded off evil spirits, scarabs of immortality, use of wax images, formulas and spells, the secret name, much more. 253pp. 5⅜ × 8½. 22681-6 Pa. $4.50

THE DANCE OF SIVA, Ananda Coomaraswamy. Preeminent authority unfolds the vast metaphysic of India: the revelation of her art, conception of the universe, social organization, etc. 27 reproductions of art masterpieces. 192pp. 5⅜ × 8½.
24817-8 Pa. $5.95

CHRISTMAS CUSTOMS AND TRADITIONS, Clement A. Miles. Origin, evolution, significance of religious, secular practices. Caroling, gifts, yule logs, much more. Full, scholarly yet fascinating; non-sectarian. 400pp. 5⅜ × 8½.
23354-5 Pa. $6.95

THE HUMAN FIGURE IN MOTION, Eadweard Muybridge. More than 4,500 stopped-action photos, in action series, showing undraped men, women, children jumping, lying down, throwing, sitting, wrestling, carrying, etc. 390pp. 7⅞ × 10⅝.
20204-6 Cloth. $24.95

THE MAN WHO WAS THURSDAY, Gilbert Keith Chesterton. Witty, fast-paced novel about a club of anarchists in turn-of-the-century London. Brilliant social, religious, philosophical speculations. 128pp. 5⅜ × 8½. 25121-7 Pa. $3.95

A CEZANNE SKETCHBOOK: Figures, Portraits, Landscapes and Still Lifes, Paul Cezanne. Great artist experiments with tonal effects, light, mass, other qualities in over 100 drawings. A revealing view of developing master painter, precursor of Cubism. 102 black-and-white illustrations. 144pp. 8⅜ × 6⅜. 24790-2 Pa. $6.95

AN ENCYCLOPEDIA OF BATTLES: Accounts of Over 1,560 Battles from 1479 B.C. to the Present, David Eggenberger. Presents essential details of every major battle in recorded history, from the first battle of Megiddo in 1479 B.C. to Grenada in 1984. List of Battle Maps. New Appendix covering the years 1967–1984. Index. 99 illustrations. 544pp. 6½ × 9¼. 24913-1 Pa. $14.95

AN ETYMOLOGICAL DICTIONARY OF MODERN ENGLISH, Ernest Weekley. Richest, fullest work, by foremost British lexicographer. Detailed word histories. Inexhaustible. Total of 856pp. 6½ × 9¼.
21873-2, 21874-0 Pa., Two-vol. set $19.90

WEBSTER'S AMERICAN MILITARY BIOGRAPHIES, edited by Robert McHenry. Over 1,000 figures who shaped 3 centuries of American military history. Detailed biographies of Nathan Hale, Douglas MacArthur, Mary Hallaren, others. Chronologies of engagements, more. Introduction. Addenda. 1,033 entries in alphabetical order. xi + 548pp. 6½ × 9¼. (Available in U.S. only)
24758-9 Pa. $13.95

LIFE IN ANCIENT EGYPT, Adolf Erman. Detailed older account, with much not in more recent books: domestic life, religion, magic, medicine, commerce, and whatever else needed for complete picture. Many illustrations. 597pp. 5⅜ × 8½.
22632-8 Pa. $8.95

HISTORIC COSTUME IN PICTURES, Braun & Schneider. Over 1,450 costumed figures shown, covering a wide variety of peoples: kings, emperors, nobles, priests, servants, soldiers, scholars, townsfolk, peasants, merchants, courtiers, cavaliers, and more. 256pp. 8⅜ × 11¼. 23150-X Pa. $9.95

THE NOTEBOOKS OF LEONARDO DA VINCI, edited by J. P. Richter. Extracts from manuscripts reveal great genius; on painting, sculpture, anatomy, sciences, geography, etc. Both Italian and English. 186 ms. pages reproduced, plus 500 additional drawings, including studies for *Last Supper, Sforza* monument, etc. 860pp. 7⅞ × 10¾. (Available in U.S. only) 22572-0, 22573-9 Pa., Two-vol. set $31.90

CATALOG OF DOVER BOOKS

AMERICAN CLIPPER SHIPS: 1833–1858, Octavius T. Howe & Frederick C. Matthews. Fully-illustrated, encyclopedic review of 352 clipper ships from the period of America's greatest maritime supremacy. Introduction. 109 halftones. 5 black-and-white line illustrations. Index. Total of 928pp. 5⅜ × 8½.
25115-2, 25116-0 Pa., Two-vol. set $17.90

TOWARDS A NEW ARCHITECTURE, Le Corbusier. Pioneering manifesto by great architect, near legendary founder of "International School." Technical and aesthetic theories, views on industry, economics, relation of form to function, "mass-production spirit," much more. Profusely illustrated. Unabridged translation of 13th French edition. Introduction by Frederick Etchells. 320pp. 6⅛ × 9¼. (Available in U.S. only) 25023-7 Pa. $8.95

THE BOOK OF KELLS, edited by Blanche Cirker. Inexpensive collection of 32 full-color, full-page plates from the greatest illuminated manuscript of the Middle Ages, painstakingly reproduced from rare facsimile edition. Publisher's Note. Captions. 32pp. 9⅜ × 12¼. 24345-1 Pa. $4.95

BEST SCIENCE FICTION STORIES OF H. G. WELLS, H. G. Wells. Full novel *The Invisible Man,* plus 17 short stories: "The Crystal Egg," "Aepyornis Island," "The Strange Orchid," etc. 303pp. 5⅜ × 8½. (Available in U.S. only)
21531-8 Pa. $6.95

AMERICAN SAILING SHIPS: Their Plans and History, Charles G. Davis. Photos, construction details of schooners, frigates, clippers, other sailcraft of 18th to early 20th centuries—plus entertaining discourse on design, rigging, nautical lore, much more. 137 black-and-white illustrations. 240pp. 6⅛ × 9¼.
24658-2 Pa. $6.95

ENTERTAINING MATHEMATICAL PUZZLES, Martin Gardner. Selection of author's favorite conundrums involving arithmetic, money, speed, etc., with lively commentary. Complete solutions. 112pp. 5⅜ × 8½. 25211-6 Pa. $2.95

THE WILL TO BELIEVE, HUMAN IMMORTALITY, William James. Two books bound together. Effect of irrational on logical, and arguments for human immortality. 402pp. 5⅜ × 8½. 20291-7 Pa. $7.95

THE HAUNTED MONASTERY and THE CHINESE MAZE MURDERS, Robert Van Gulik. 2 full novels by Van Gulik continue adventures of Judge Dee and his companions. An evil Taoist monastery, seemingly supernatural events; overgrown topiary maze that hides strange crimes. Set in 7th-century China. 27 illustrations. 328pp. 5⅜ × 8½. 23502-5 Pa. $6.95

CELEBRATED CASES OF JUDGE DEE (DEE GOONG AN), translated by Robert Van Gulik. Authentic 18th-century Chinese detective novel; Dee and associates solve three interlocked cases. Led to Van Gulik's own stories with same characters. Extensive introduction. 9 illustrations. 237pp. 5⅜ × 8½.
23337-5 Pa. $5.95

Prices subject to change without notice.

Available at your book dealer or write for free catalog to Dept. GI, Dover Publications, Inc., 31 East 2nd St., Mineola, N.Y. 11501. Dover publishes more than 175 books each year on science, elementary and advanced mathematics, biology, music, art, literary history, social sciences and other areas.